OBJECTIVE QUESTIONS IN PETROLEUM ENGINEERING

(Important Multiple Choice Questions with Answers)

2nd Edition

Dr. Vikas Mahto
Associate Professor
Department of Petroleum Engineering
Indian School of Mines, Dhanbad-826004 (India)

KHANNA BOOK PUBLISHING CO. (P) LTD.
Publisher of Science, Technology and Engineering Books
4C/4344, Ansari Road, Darya Ganj, New Delhi-110002
Phone: 011-23244447-48 **Mobile:** +91-99109 09320
E-mail: contact@khannabooks.com
Website: www.khannabooks.com

Price : ₹ 245.00

Objective Questions in Petroleum Engineering
Dr. Vikas Mahto

ISBN: 978-93-86173-24-9

Edition: 2016
Reprint: 2021

Published by:

KHANNA BOOK PUBLISHING CO. (P) LTD.
4C/4344, Ansari Road, Darya Ganj, New Delhi-110 002
Phone: 011-23244447-48 **Mobile:** +91-9910909320
E-mail: contact@khannabooks.com

Typeset and Cover Design by:
Book One Graphics, New Delhi

Printed in India by:
India Book Printers & Binders, Delhi

Preface

This book contains numerous important questions which have been methodically prepared/selected from different text books, manuals of petroleum industries, SPE technical papers and teaching materials of distinguished persons. These questions are very relevant for promoting fundamental understanding of petroleum engineering. The various topics covered in this book are petroleum exploration, oil & gas well drilling, reservoir engineering, offshore oil & gas operations, petroleum production engineering, oil & gas well testing, enhanced oil recovery techniques, health, safety & environment in petroleum industries, and latest trends in petroleum industry. It contains objective type theoretical and numerical questions based on basic concepts & latest trends in the petroleum exploration and production industries.

This book will be primarily useful for fresh graduates of petroleum engineering who can prepare themselves soundly for both written as well as oral examinations. At the same time this can be used as readily available handbook for quick reference by practicing engineers. There is no other comprehensive book available of this format in petroleum engineering discipline which can build basic concepts of all the important courses of petroleum engineering.

I would like to acknowledge the head of the department for his kind support. I also wish to acknowledge the contribution of Mr. Rajat Jain, Senior Research Fellow, who has helped in typing the contents of this book.

I shall greatly appreciate being informed of suggestions for the improvement of the book.

VIKAS MAHTO

About the Author

Dr Vikas Mahto is an Associate Professor in the Petroleum Engineering Department of Indian School of Mines Dhanbad, India. He has received his Ph.D. Degree in Petroleum Engineering from Indian School of Mines Dhanbad in 2004. He is having more than 12 Years of Experience in the field of Chemical Engineering and Petroleum Engineering. He is the author of more than 100 technical papers in different national/international journals & conferences of repute. He has completed three research projects sponsored by UGC-ISM, UGC New Delhi and CSIR New Delhi and currently working on collaborative R&D project with Oil India Ltd, Duliajan, Assam. He has filed two Indian patents in his name. Four students have awarded their Ph.D. Degree in petroleum Engineering under his supervision and eight students are pursuing their Ph,D. work under his guidance. He is the reviewer and member of editorial boards of many national and international journals.

Contents

CHAPTER 1

Petroleum Exploration

1. Which of the following rock favors petroleum formation?
   a) Sedimentary rock
   b) Igneous rock
   c) Metamorphic rock
   d) None of the above

2. Which of the following is a type of sedimentary rock?
   a) Organic rock
   b) Chemical rock
   c) Clastic rock
   d) All of the above

3. The set of processes transforming sediment into rocks is called
   a) Lithification
   b) Hydrolysis
   c) Cementing
   d) None of the above

4. The formation of bed created by wind that changes direction and the layers meet at different angles is known as
   a) Cross bedding
   b) Graded bedding
   c) Both (a) and (b)
   d) None of the above

5. The formation of horizontal or nearly horizontal layers of bed at the time of deposition, in which the coarsest particles are concentrated at the bottom and grade gradually upward into fine silt is known as
   a) Cross bedding
   b) Graded bedding
   c) Both (a) and (b)
   d) None of the above

6. A source rock that is too immature to generate petroleum in its natural setting is known as
   a) Potential source rock
   b) Effective source rock
   c) Both (a) and (b)
   d) None of the above

7. A source rock that has already formed and expelled petroleum to a reservoir is known as
   a) Potential source rock
   b) Effective source rock
   c) Both (a) and (b)
   d) None of the above

8. Which of the following is most abundant sedimentary rock found in the earths's crust?
   a) Sandstone
   b) Shale
   c) Limestone
   d) None of the above

9. Which of the following is not among the basic type of organic matters in sediments?
   a) Type I
   b) Type III
   c) Type V
   d) Type IV

10. Which of the following type of organic matters in sediments include ancient oil shales of marine origin?
   a) Type I
   b) Type II
   c) Type III
   d) Type IV

11. _______ includes the principle source sediments for oil.
   a) Type I
   b) Type II
   c) Type III
   d) Type IV

12. Which of the following is a important method for geochemical exploration of oil?
   a) Geothermal survey
   b) Radioactivity Survey
   c) Both (a) and (b)
   d) None of the above

13. Oil and gas are found in _______
   a) Shale
   b) Sand
   c) Granite
   d) Sulfur

14. Reservoir rock containing petroleum has
   a) Low permeability
   b) High permeability
   c) High porosity
   d) Both (b) and (c)

15. Which of the following is true about primary porosity?
   a) Remains in place after deposition and pores are connected
   b) Part of the rock is dissolved after deposition.
   c) Pores are not connected
   d) None of the above

16. Which of the following is true about secondary porosity?
   a) Part of the rock is dissolved after deposition and pores are not connected
   b) Remains in place after deposition and pores are connected
   c) Both (a) and (b)
   d) None of the above

17. Which of the following is a type of primary porosity?
   a) Intercrystalline porosity
   b) Moldic Porosity
   c) Intraparticle porosity
   d) None of the above

18. Which of the following is a type of secondary porosity?
   a) Intergrannular porosity
   b) Intraparticle porosity
   c) Intercrystalline porosity
   d) None of the above

19. Which of the following statement is true?
    a) Permeability increases with grain size
    b) Permeability increases as sorting becomes poorer
    c) Tight packing favours lesser permeability
    d) None of the above

20. Which of the following statement is false?
    a) Porosity decreases as sorting becomes poorer
    b) Tighter packing favours more porosity
    c) Permeability decreases with grain size
    d) None of the above

21. In general, fine grained rocks have
    a) High porosity
    b) High Permeability
    c) High porosity and permeability
    d) High porosity and low permeability

22. Which of the following materials has the highest porosity
    a) Clay
    b) Silt
    c) Gravels
    d) Sandstone

23. A source rock ________
    a) Reflects high productivity
    b) Is igneous
    c) Is usually a sandstone with impermeable layers above and below
    d) Is created by plate tectonics

24. Which of the following is not possible in oil trap?
    a) Anticline
    b) Lake
    c) Fold
    d) Fault

25. Migration of oil can occur in_____
    a) Shale
    b) An oil pool
    c) Pore spaces and fractures
    d) Limestone

26. A reservoir rock ______
    a) Reflects high productivity
    b) Is igneous
    c) Is usually sandstone with impermeable layers above and below
    d) Is created by plate tectonics

27. The hydrocarbon fuel in the oil shale is:
    a) Petroleum
    b) Kerogen
    c) Natural gas
    d) methane

28. When rocks near the surface break, the two halves might move in relation to each other, producing a fracture is called
    a) Trap
    b) Fault
    c) Fold
    d) None of the above

29. The cap rock is________
    a) Layer of impermeable rock above the reservoir
    b) Layer of impermeable rock below the reservoir
    c) Layer of permeable rock below the reservoir
    d) Layer of permeable rock above the reservoir

30. ____________ states that oil and gas come predominantly from the remains of microscopic plants and small animal organisms.
    a) Organic theory
    b) Inorganic theory
    c) Biogenic theory
    d) None of the above

31. The arrangement of rock layers containing an accumulation of hydrocarbons where the formation prevents the hydrocarbons from rising to the surface is called
    a) Trap
    b) Fault
    c) Fold
    d) Cap

32. Those in which porosity or permeability has changed within a formation or where a nonporous layer seals of the top of the reservoir are known as
    a) Structural trap
    b) Stratigraphic trap
    c) Anticline
    d) Syncline

33. ______ are kerogen rich rocks which are hot enough to generate oil.
    a) Source rocks
    b) Reservoir rocks
    c) Both (a) and (b)
    d) None of the above

34. In __________, rock layers that were originally formed horizontally are folded upward into an arc or dome.
    a) Syncline trap
    b) Stratigraphic trap
    c) Anticlinal trap
    d) None of the above

35. Rock strata that have crumbled and buckled into wave like structures are called
    a) Faults
    b) Traps
    c) Folds
    d) None of the above

36. A film of water that sticks to the solid rock materials surrounding the pore spaces is called
    a) Non-wetting water
    b) Produced water
    c) Connate water
    d) Wetting water

37. Long, narrow block of crust between two faults that has sunk relative to the surrounding crust is called
    a) Graben
    b) Basin
    c) Both of these
    d) None of the above

38. The movement of hydrocarbons from the source rock into permeable carrier beds is called
    a) Secondary migration
    b) Primary migration
    c) Tertiary migration
    d) None of the above

39. The oil window temperature range for hydrocarbon is ______
    a) 60-120°C
    b) 120-225°C
    c) Both (a) and (b)
    d) None of the above

40. The gas window temperature range for hydrocarbon is ______
    a) 60-120°C
    b) 120-225°C
    c) Both (a) and (b)
    d) None of the above

41. In ________, oil generation occurs between 60-120°C, and gas generation between 120-225°C.
    a) Immature kerogen
    b) Mature kerogen
    c) Gas window
    d) Oil window

42. Anticlinal traps formed by crustal tension and then compacting sedimentary layers are called:
    a) Compactional Anticlines
    b) Expandable Anticlines
    c) Non-expandable Anticlines
    d) None of the above

43. The study of modern sediments such as sand, silt, and clay, and understanding the processes that deposit them is called:
    a) Sedimentology
    b) Petrography
    c) Both (a) and (b)
    d) None of the above

44. __________ generally contain more than 1500 ppm organic carbon, but yield only a small percentage of their contained hydrocarbons.
    a) Source rocks
    b) Reservoir rocks
    c) Cap rock
    d) None of the above

45. A branch of petrology which focuses on detailed descriptions of rocks is called:
    a) Stratigraphy
    b) Petrography
    c) Both (a) and (b)
    d) None of the above

46. Zones where crustal tension causes plates to pull apart is:
    a) Divergent plate boundary
    b) Convergent plate boundary
    c) Both of these
    d) None of the above

47. Which of the following statement is true?
    a) Kaolinitic clays harm permeability more than illitic clays.
    b) Kaolinitic clays harm permeability less than illitic clays.

c) Both clays don not harm the permeability of the rocks.

d) None of the above

48. The plot of burial depth versus geologic time for a particular region is called

a) Burial History Curve

b) Burial depth curve

c) Both of these

d) None of the above

49. Gap in geologic time scale of stratigraphy is called:

a) Conformity

b) Unconformity

c) Both (a) and (b)

d) None of the above

50. Which of the following map shows elevation of the Earth's surface?

a) Base Map

b) Geologic map

c) Topographic map

d) None of the above

51. Which of the following map is used to describe subsurface rocks?

a) Structural map

b) Isopach map

c) Both (a) and (b)

d) None of the above

52. Which of the following map shows location of all wells?

a) Base Map

b) Geologic map

c) Topographic map

d) None of the above

53. Which of the following map net pay thickness of petroleum reservoir?

a) Isopach map

b) Isochore map

c) Both of (a) and (b)

d) None of the above

54. Which of the following map shows variation in true thickness of rock?

a) Isopach map

b) Isochore map

c) Both of (a) and (b)

d) None of the above

55. Which of the following map shows variation in vertical thickness of rock?

a) Isochore map

b) Isopach map

c) Both of (a) and (b)

d) None of the above

56. Which of the following statement is true?

a) Isochore maps show greater thickness than the isopach map of the same formation

b) For horizontal beds an isopach map is identical to the isochore map

c) Isopach map and isochore map differ from each other for inclined beds

d) All of the above

57. Isopay map is prepared by obtaining difference in elevation between

a) Gas-water contact and the top of the hydrocarbon bearing zone

b) Oil-water contact and the top of the hydrocarbon bearing zone

c) Both (a) and (b)

d) None of the above

58. A fracture along which rock blocks move relative to each other parallel to the fracture is called:

a) Fault

b) Trap

c) Cap

d) None of the above

59. Which of the following statement is true in case of seismic impedance?

a) It is given by seismic velocity x density

b) It determines the amplitude of reflected seismic waves of the reflector

c) None of the above

d) Both (a) and (b)

60. Which types of sedimentary rock is the primary interest for reservoir?

a) Clastic rock

b) Organic rock

c) Both (a) and (b)

d) None of the above

61. Seismic data at sea is acquired using:

a) Hydrophones

b) Air guns

c) Both (a) and (b)

d) None of the above

62. The movement of hydrocarbons through the carrier beds to the reservoir is known as

a) Primary migration

b) Secondary migration

c) Tertiary migration

d) None of the above

63. Chemically, kerogen consists of carbon, hydrogen, and oxygen with minor amounts of _______ and ______.

a) Nitrogen & Sulfur

b) Sulfur & Oxygen

c) Both (a) and (b)

d) None of the above

64. Which of the following force causes the petroleum to migrate from the source rock through the carrier beds to the reservoir rock?

a) Capillary force

b) Buoyancy force

c) Hydrodynamic force

d) All of the above

65. Which of the following property is important in determining whether a trap contains oil, gas or both?

a) Level of maturation

b) Source rock chemistry

c) Pressure and Temperature

d) All of the above

66. Oil generation occurs when
   a) Level of organic maturation(LOM) is 7-13
   b) Level of organic maturation(LOM) is 13-18
   c) Both (a) and (b)
   d) None of the above

67. Gas generation occurs when
   a) Level of organic maturation (LOM) is 7-13
   b) Level of organic maturation (LOM) is 13-18
   c) Both (a) and (b)
   d) None of the above

68. Which of the following statement (s) is true about petroleum system:
   a) It is the integration of petroleum migration with the thermal and tectonic evolution of a sedimentary basin.
   b) Vol of oil generated = Basin area x Average total thickness of source rock x Transformation ratio.
   c) Both (a) and (b)
   d) None of the above

69. Which of the following statement (s) is true about vitrine.
   a) It is one of the primary components of coals and most sedimentary kerogens.
   b) It is a type of macerals, where macerals are organic components of coal analogous to the minerals of rocks.
   c) Both (a) and (b)
   d) None of the above

70. Which of the following statement is true about connate water?
   a) The interstitial water existing in the reservoir rock prior to disturbance by drilling, and waters which have been buried in a closed hydraulic system and have not formed part of the hydraulic cycle for a long time.
   b) It may contain more dissolved solids, a lower percentage of sulfates, magnesium, and often calcium, and a higher percentage of sodium, potassium and chlorides than seawater.
   c) Both (a) and (b)
   d) None of the above

71. In Overmature Kerogen:
   a) Neither oil nor gas remains
   b) Only gas remains
   c) Only oil remains
   d) None of the above

72. Which of the following statement is true about the 4D Seismic Survey?
   a) It is Time-lapse 3D seismic technology.
   b) It is the use of 3D seismic surveys acquired at different times in the productive life of a reservoir.
   c) Both (a) and (b)
   d) None of the above

73. Which of the following statement is false?

a) Geophysical Exploration uses surface methods to measure the physical properties of the subsurface Earth.

b) It does not detect the presence and concentrations of ore minerals

c) It detects the presence and concentrations of hydrocarbons.

d) None of the above

74. Which of the following statement is true about the Bottom Simulating Reflector?

a) It is the label of the lowest boundary of the stable gas-hydrate prone zone.

b) Below the BSR methane occurs as pore filling gas.

c) Both (a) and (b)

d) None of the above

75. Which of the following statement is true about the abiogenic theory?

a) The idea of abiogenic petroleum origin proposes that large amounts of carbon exist naturally in the planet, some in the form of hydrocarbons.

b) Hydrocarbons are less dense than aqueous pore fluids, and migrate upward through deep fracture networks

c) Both (a) and (b)

d) None of the above

76. Which of the following statement (s) is true about gravity surveying?

a) It is based on Newton's Law of gravity

b) Free air anomaly is the variation in G due to elevation of the measuring station.

c) Bouguer anomaly is the variation in G due to mass of rock between the survey station and a reference datum, typically the sea level.

d) All of the above

77. The study of prehistoric life, including organisms' evolution and interactions with each other and their environments is called

a) Paleontology

b) Petrology

c) Sedimentology

d) All of the above

78. The initial stage of kerogen alteration is termed as

a) Catagenesis

b) Metagenesis

c) Diagenesis

d) None of the above

79. The process where kerogen is subjected to more heat, and is further changed into liquid and gaseous hydrocarbons

a) Catagenesis

b) Metagenesis

c) Diagenesis

d) None of the above

80. Which type of Kerogen has potential for mainly oil?
   a) Kerogen I
   b) Kerogen IIA
   c) Kerogen IIB
   d) Kerogen III

81. Which type of Kerogen has potential for mainly gas?
   a) Kerogen I
   b) Kerogen IIA
   c) Kerogen IIB
   d) Kerogen III

82. Which type of Kerogen has potential for oil and gas both?
   a) Kerogen I
   b) Kerogen IIA
   c) Kerogen IIB
   d) Kerogen III

83. The H/C Ratio for Kerogen III is
   a) 0.5-0.8
   b) 0.8-1.3
   c) More than 1,5
   d) None of the above

84. The H/C Ratio for Kerogen I is
   a) 0.5-0.8
   b) 0.8-1.3
   c) More than 1,5
   d) None of the above

85. The H/C Ratio for Kerogen IIB is
   a) 0.5-0.8
   b) 0.8-1.3
   c) More than 1,5
   d) None of the above

86. Which of the following is true about Vitrinite Reflectance
   a) It is a standard method for the measurement of organic saturation
   b) It uses a light microscope
   c) Both (a) and (b)
   d) None of the above

87. The arrangement of rock layers containing an accumulation of hydrocarbons where the formation prevents the hydrocarbons from rising to the surface is called
   a) Trap
   b) Fault
   c) Fold
   d) None of the above

88. In ________, porosity or permeability has changed within a formation or where a nonporous layer seals of the top of the reservoir.
   a) Statigraphic trap
   b) Structural trap
   c) Both (a) and (b)
   d) None of the above

89. Trap formed by the folding or faulting of reservoir rock is called
   a) Structural trap
   b) Stratigraphic trap
   c) Fault trap
   d) None of the above

90. Trap formed by the movement of rock along a fault rock is known as
   a) Fault trap

b) Structural Trap

c) Stratigraphic trap

d) None of the above

91. A two-dimensional slice from the surface of the earth downward is known as

a) Seismic section

b) Geological section

c) Basin section

d) None of the above

92. ___________ are the sensors that pick up the reflected seismic waves and send them through cables to a recorder.

a) Geophones

b) Reflector

c) Refractor

d) None of the above

93. Which of the following is true about geophone?

a) It transforms seismic energy into an electrical voltage

b) It converts change of pressure to an electrical voltage

c) Piezoelectric crystal is used as a sensing element

d) None of the above

94. Hydrophones are

a) Onshore version of geophones

b) Marine version of geophones

c) Both (a) and (b)

d) None of the above

95. Which of the following is true about the hydrophone?

a) It is a electromechanical transducer

b) It transforms seismic energy into an electrical voltage

c) Piezoelectric crystal is used as a sensing element

d) None of the above

96. Which of the following is most common in marine 3D operation?

a) T spread

b) Cross spread

c) Parallel CMP profiles

d) None of the above

97. Which of the following technique is used for position location in marine shooting?

a) Radio positioning

b) Satellite navigation

c) Both (a) and (b)

d) None of the above

98. Theodolites are used for

a) Marine shooting

b) Land survey

c) Both (a) and (b)

d) None of the above

99. Global positioning system is used for

a) Marine shooting

b) Land shooting

c) Both (a) and (b)

d) None of the above

100. Vertical seismic profiling is used for

a) Identification of acoustic boundaries

b) Correlation with logs and existing seismic lines

c) Prediction of impedance below the well bottom

d) All of the above

101. Remote sensing is used for

a) Identification of surface indications of hydrocarbons

b) Structural mapping and rock type mapping

c) Geomorphic analysis

d) All of the above

102. Which of the following method measures spatial variations in the earths ?

a) Seismic reflection survey

b) Gravity survey

c) Magnetic Survey

d) None of the above

103. Which of the following method investigates subsurface geology by measuring the intensity of its magnetic fields?

a) Seismic reflection survey

b) Gravity survey

c) Magnetic Survey

d) None of the above

104. The trap, where hydrocarbon is trapped due to hydrodynamic flow of the water in permeable beds is known as

a) Hydrodynamic trap

b) Stratigraphic trap

c) Structural trap

d) Anticlinal trap

105. The ratio of the pore volume to the bulk volume of a material is called

a) Porosity

b) Permeability

c) Both (a) and (b)

d) None of the above

106. __________ is the measure of pore space filled with water.

a) Water saturation

b) Porosity

c) Permeability

d) None of the above

107. __________is a measure of measure of free space present in a rock that can potentially be occupied by hydrocarbons

a) Porosity

b) Permeability

c) Both (a) and (b)

d) None of the above

108. Which of the following describes about the total of all pores in the rock?

a) Total porosity

b) Effective porosity

c) Both (a) and (b)

d) None of the above

109. Which of the following describes about the pores that are interconnected and can be used to store hydrocarbons?

a) Effective porosity

b) Absolute porosity

c) Both (a) and (b)

d) None of the above

110. ________ is a measure of attraction between rock and fluids.

a) Wettability

b) Porosity

c) Saturation

d) None of the above

111. Which of the following pairs of properties best describes the characteristics of rock layer in which the oil and gas are found?

a) Porous and Permeable

b) Porous and Impermeable

c) Nonporous and Permeable

d) Nonporous and Impermeable

112. Geomagnetic secular variations originate from the________.

a) Inner core

b) Outer Core

c) Crust

d) Mantle

113. Which of the following is similar to removal of regional compound from magnetic data?

a) Band Pass Filtering

b) Low Pass Filtering

c) High Pass Filtering

d) Band Reject Filtering

114. Which of the following is the most common hydrocarbon indicator?

a) Flat spot

b) Dim spot

c) Bright spot

d) Velocity spot

115. Which of the following hydrocarbon is found in oil shale?

a) Petroleum

b) Kerogen

c) Natural Gas

d) Methane

116. Which of the following is most sensitive instrument for magnetic survey?

a) Magnetic Field Balance

b) Flux Gate Magnetometer

c) Proton Precision Magnetometer

d) Optically Pumped Magnetometer

117. Which of the following basins is producing petroleum from coal rich reservoir rocks?

a) Rajasthan Basin

b) Cambay Basin

c) Cauvery Basin

d) Krishna Godavari Basin

118. Which of the following country consumes more energy?

a) USA

b) China

c) India

d) Germany

119. All oil traps contain

a) An impermeable layer

b) Anticline

c) Fault

d) all of the above

120. Which of the following rock types would most likely be the best oil reservoir?
    a) Granite
    b) Shale
    c) Sandstone
    d) Salt

121. The hydrocarbon derived from coal, oil shales or tar sand is called
    a) Natural Gas
    b) Biomass
    c) Syncrude
    d) None of the above

122. Oil, natural gas, and coal are examples of
    a) Fossil fuels.
    b) Hydrocarbon fuel.
    c) Nonrenewable resources.
    d) All of the above

123. As societies evolve from primitive to technological, the energy consumption associated with agriculture________
    a) Increases.
    b) Decreases
    c) Remains Constant
    d) May either increase or decrease depending upon what foods the individuals eat

124. As heat and pressure act upon petroleum, it finally becomes______
    a) Oil.
    b) Tar.
    c) Natural gas.
    d) Coal.

125. One barrel of oil is equal to how many gallons?
    a) 21
    b) 42
    c) 56
    d) 100

126. The nation having the largest proven reserves of crude oil is
    a) United States.
    b) Kuwait.
    c) Saudi Arabia.
    d) Algeria.

127. Petroleum accounts for about percent of the energy used in the United States.
    a) 10
    b) 25
    c) 40
    d) 75

128. Natural gas supplies about _percent of the energy used in the United States.
    a) 5
    b) 11
    c) 15
    d) 25

129. An oil well gusher would be an example of
    a) Primary recovery.
    b) Secondary recovery.
    c) Tertiary recovery.
    d) Enhanced recovery.

130. Geopressurized zones might one day become an important source of

a) Petroleum.

b) Natural gas.

c) Lignite.

d) Kerosene

131. The reservoir now believed to contain the largest amount of carbon is

a) Atmosphere.

b) Natural gas.

c) Coal (bituminous and anthracite).

d) Gas hydrates.

132. Which one of the following fuels would give off the most heat per given unit weight?

a) Peat

b) Lignite

c) Bituminous

d) Anthracite

133. Coal subjected to overly high temperatures will convert metamorphose into

a) Graphite.

b) Oil shale.

c) Tar sand.

d) Peat.

134. The largest recoverable U.S. energy reserve is

a) Petroleum.

b) Coal.

c) Natural gas.

d) Uranium oxide.

135. The worst oil spill in United States waters was the

a) Santa barbara spill

b) Amoco cadiz spill

c) Exxon valdez spill

d) Torrey canyon spill

136. Coal supplies about _percent of the energy used in the United States.

a) 10

b) 25

c) 50

d) 80

137. The substance found in coal that contributes to acid rainfall is

a) Carbon.

b) Sulfur.

c) Chlorine.

d) Nitrogen.

138. The environmental disaster that has affected the town of Centralia, Pennsylvania was caused by

a) Toxic coal-ash residue.

b) Acid rainfall.

c) An underground coal-mine fire.

d) An oil spill.

139. The United States has about two-thirds of the world's known resources of

a) Tar sands.

b) Natural gas.

c) Coal.

d) Oil shale

140. All of the following are problems in the development of oil shale in the U.S. as a viable energy resource except
   a) Extraction and processing technologies are not competitive with those of conventional petroleum.
   b) Environmental problems associated with the surface mining of oil shale.
   c) The problem of waste rock disposal.
   d) The low quality of the oil

141. The three dimensional orientation of beds is described by
   a) Dip of the bedding plane
   b) Strike of the bedding plane
   c) Both (a) and (b)
   d) None of the above

142. The structural relationship between unconformity and the underlying rock is known as
   a) Overstep
   b) Overlap
   c) Both (a) and (b)
   d) None of the above

143. The structural relationship between unconformity and the overlying rock is known as
   a) Overstep
   b) Overlap
   c) Both (a) and (b)
   d) None of the above

144. The point of maximum curvature of a folded bed is called______
   a) Hinge
   b) Limb
   c) Both (a) and (b)
   d) None of the above

145. When a fold is convex up , it is called an___
   a) Anticline
   b) Syncline
   c) Both (a) and (b)
   d) None of the above

146. When a fold is convex down, it is called an___
   a) Anticline
   b) Syncline
   c) Both (a) and (b)
   d) None of the above

147. Which of the following is the example of anticline?
   a) Dome
   b) Basin
   c) Both (a) and (b)
   d) None of the above

148. Which of the following is the example of syncline?
   a) Dome
   b) Basin
   c) Both (a) and (b)
   d) None of the above

149. A fault that shows relative downward displacement of hanging wall is termed as
   a) Normal Fault
   b) Reverse Fault
   c) Stirike Slip Fault
   d) None of the above

150. A fault that shows relative upward displacement of hanging wall is termed as

a) Normal Fault

b) Reverse Fault

c) Stirike slip fault

d) None of the above

151. A fault that shows sidewise displacement of hanging wall is termed as

a) Normal Fault

b) Reverse Fault

c) Strike Slip Fault

d) None of the above

152. A down thrown block bound by two normal faults inclined towards each other is called

a) Graben

b) Horst

c) Both (a) and (b)

d) None of the above

153. A up thrown block bound by two normal faults inclined towards each other is called

a) Graben

b) Horst

c) Both (a) and (b)

d) None of the above

Answers									
Q1.	a	Q36.	d	Q71.	a	Q106.	a	Q141.	c
Q2.	d	Q37.	a	Q72.	c	Q107.	a	Q142.	a
Q3.	a	Q38.	b	Q73.	b	Q108.	a	Q143.	b
Q4.	a	Q39.	a	Q74.	c	Q109.	a	Q144.	a
Q5.	b	Q40.	a	Q75.	c	Q110.	a	Q145.	a
Q6.	a	Q41.	b	Q76.	d	Q111.	a	Q146.	b
Q7.	b	Q42.	a	Q77.	a	Q112.	b	Q147.	a
Q8.	b	Q43.	a	Q78.	c	Q113.	c	Q148.	b
Q9.	c	Q44.	a	Q79.	a	Q114.	c	Q149.	a
Q10.	b	Q45.	b	Q80.	a	Q115.	b	Q150.	b
Q11.	b	Q46.	a	Q81.	d	Q116.	b	Q151.	c
Q12.	c	Q47.	a	Q82.	c	Q117.	b	Q152.	a
Q13.	a	Q48.	a	Q83.	a	Q118.	b	Q153.	b
Q14.	d	Q49.	b	Q84.	c	Q119.	a		
Q15.	a	Q50.	c	Q85.	b	Q120.	c		
Q16.	a	Q51.	c	Q86.	c	Q121.	c		
Q17.	c	Q52.	a	Q87.	a	Q122.	d		
Q18.	c	Q53.	a	Q88.	a	Q123.	a		
Q19.	c	Q54.	a	Q89.	a	Q124.	c		
Q20.	b	Q55.	a	Q90.	a	Q125.	b		
Q21.	c	Q56.	d	Q91.	a	Q126.	c		
Q22.	a	Q57.	c	Q92.	a	Q127.	c		
Q23.	a	Q58.	a	Q93.	a	Q128.	d		
Q24.	b	Q59.	d	Q94.	b	Q129.	a		
Q25.	c	Q60.	c	Q95.	c	Q130.	b		
Q26.	c	Q61.	c	Q96.	c	Q131.	d		
Q27.	b	Q62.	b	Q97.	c	Q132.	d		
Q28.	b	Q63.	a	Q98.	b	Q133.	a		
Q29.	a	Q64.	d	Q99.	c	Q134.	b		
Q30.	a	Q65.	d	Q100.	d	Q135.	c		
Q31.	a	Q66.	a	Q101.	d	Q136.	b		
Q32.	b	Q67.	b	Q102.	b	Q137.	b		
Q33.	a	Q68.	c	Q103.	c	Q138.	c		
Q34.	c	Q69.	c	Q104.	a	Q139.	d		
Q35.	c	Q70.	c	Q105.	a	Q140.	d		

CHAPTER 2

Oil and Gas Well Drilling Technology

1. What is the objective of well planning?
   a) It provides a comprehensive assessment of a drilling project
   b) It formulate a program from many requisite variables for safe drilling with minimum cost
   c) It is aimed at achieving a safe and economic delivery of the project
   d) All of the above

2. Which of the following is the component of a well planning report?
   a) Geological Informations
   b) Rig selection, Drill String Design and Casing Design
   c) Mud program and Cement Program
   d) All of the above

3. Crown Block and Travelling Block are part of –
   a) Drilling rig
   b) Drilling bit
   c) Drilling fluid
   d) Drilling cost

4. What is the factor on which drilling cost depends?
   a) drilling location
   b) drilling time
   c) materials and services cost
   d) All of the above

5. Which of the following is the function of a drilling string?
   a) It allows fluids to circulate through the well
   b) It allows the bit to be rotated
   c) It has sufficient strength for hoisting and rotation
   d) All of the above

6. The topmost casing is
   a) Surface casing
   b) Intermediate casing
   c) Production casing
   d) Conductor casing

7. Surface casing is installed to
   a) Prevent the bit from formation fluid
   b) Protect the integrity of the hole
   c) Protect deep fresh ground water zone
   d) b &c

8. A short piece of pipe which is used to connect two different diameter drillstring component is called
   a) Centralizer
   b) Jar

c) Crossover

d) None of the above

9. Which of the following component is not a principle component of the fluid circulating system?

a) Mud pump

b) Shale shaker

c) Hydrocyclone

d) None of the above

10. Which of the following component of a drill string is used to put weight on the drill bit?

a) Float collar

b) Drilling line

c) Ball sealers

d) Drill collars

11. The hoisting system does not include

a) Draw works

b) Block and tackle

c) Rotary table

d) None of the above

12. Which of the following is the function of rotary table?

a) To rotate the drill string and bit

b) To rotate the Kelly and drill string

c) To house and rotate the master bushing

d) All of the above

13. Which of the following cementing tool keep cement separated from mud?

a) Float Collar

b) Wiper plug

c) Casing shoe

d) None of the above

14. Which of the following casing stabilizes hole and protects weak, low pressure zones from heavy mud weights?

a) Surface casing

b) Intermediate casing

c) Conductor casing

d) None of the above

15. Which of the following item is the essential part of hoisting system

a) Mud pump

b) Draw works

c) Shale Shakers

d) Hydrocyclone

16. __________ is a type of bit

a) Roller cone

b) Two cone

c) Fixed cutter

d) All of the above

17. Steel pipe placed in oil and gas well to prevent the well bore from caving in is

a) Drill string

b) Annulus

c) Casing

d) None of the above

18. What is the major function of the drill collars?

a) To provide sufficient weight on the drill bit

b) To make sure drillstring is always in tension

c) Both (a) and (b)

d) None of the above

19. The function of drilling slip is

a) To hold the drill pipe in place

b) To prevent drill pipe from slipping

c) It helps in making and breaking connection

d) All of the above

20. What is used to stablize and prevent caving of the wellbore?

a) Casing

b) Tubing

c) Stabilizer

d) None of the above

21. What is the most common range of drill pipe?

a) 10 ft

b) 27-30 ft

c) 50 ft

d) 60 ft

22. Hydrostatic pressure is --- in deep wells than in shallow wells

a) Less

b) Greater

c) Very less

d) None of the above

23. If the casing program for a well calls for 5000 feet of 4.5" casing and 2000 ft 8 5/8" casing which one is set first?

a) 8 5/8".

b) 7 6/9"

c) 8 7/8"

d) 10 8/5"

24. What type of log could be ran to determine if the primary cementing job was leaking?

a) Cement bond log

b) Gamma ray

c) Neutron porosity

d) SP log

25. What is done with the drilling mud prior to cementing the casing?

a) Conditioning

b) Perforating

c) Both (a) and (b)

d) None of the above

26. What type of rig is currently being used to drill wells from the deepest water depth in the Gulf of Mexico?

a) Spar

b) Jack up

c) Semi-submersible

d) None of the above

27. The process of drilling multiple wells near to one other is known as

a) Pad drilling

b) Spudding in

c) Both (a) and (b)

d) None of the above

28. Which of the following is true about cutting transport in deviated and horizontal wells?

a) Increasing the annular velocity increases the cuttings bed height significantly

b) Low viscosity mud is more effective in cuttings transport than high viscosity at the same flow rate

c) A small increase in mud density increases the cuttings bed height.

d) None of the above

29. Filter cake thickness is measured and reported to

a) 1/32 of an inch

b) 1/64 of an inch

c) 1/16 of an inch

d) 1/48 of an inch

30. Funnel viscosity is written in

a) Centipoises

b) Poise

c) Sec/qt

d) None of the above

31. The location of a well determines

a) Site preparing costs

b) Rig moving cost

c) Daily operating cost of the drilling operation

d) All of the above

32. Unintentional flow of formation fluid into the wellbore is called

a) Kill fluid

b) Kick

c) Light mud

d) A and C

33. Which one of the following is not an indication of a kick?

a) Decrease in pump

b) Pit gain

c) Decrease in pump stroke

d) Increase in flow rate

34. What two types of cement additives are used to control setting time?

a) Accelerators

b) Retarders

c) Both of these

d) None of the above

35. Which of the following is a specialized casing equipment used for cementing operations?

a) Wall scratchers,

b) Centralizers

c) Guide shoe

d) All of the above

36. What is the name for the longest string of casing ran in a well?

a) Surface casing

b) Intermediate casing

c) Production casing

d) Conductor casing

37. The process when the drillpipe is pulled out from the well is called

a) Tripping out

b) Tripping in

c) Surging

d) None of the above

38. The process when the drillpipe is run back into the well

a) tripping in

b) tripping out

c) Swabbing

d) None of the above

39. What is the procedure to unscrew stuck pipe below the surface?
    a) Back off operation
    b) Killing
    c) Milling
    d) None of the above

40. Which of the following statements is incorrect about blind rams?
    a) They are designed to close when no pipe is in the hole
    b) They flatten drill pipe if accidentally used
    c) They can stop the flow from the well if used when the drill string is in the hole
    d) A and B

41. ______are two large blocks of steel that are designed to close the around the pipe in the well
    a) Pipe rams
    b) Shear rams
    c) Blind rams
    d) None of the above

42. ________are designed to cut across any pipe in the well to close the well quickly
    a) Shear rams
    b) Pipe rams
    c) Blind rams
    d) None of the above

43. _______a steel, spool-shaped fitting that permits attachment of kill and chokelines to the stack
    a) Drilling stool
    b) Spool
    c) Both
    d) None of the above

44. ________are used to close off the top of the well.
    a) Blowout preventers
    b) Lubricator
    c) Swivel
    d) None of the above

45. All tubulars are specified by
    a) Range, weight per foot, outside diameter, steel composition
    b) Range, weight per foot, inner diameter, steel grade
    c) Manufacturers name, weight per foot, outside diameter, steel grade
    d) Range, Weight per foot, outside diameter, steel grade

46. Rate at which angle changes during deviation, usually ~2 degrees/ 100 ft is called
    a) Build rate
    b) Drop rate
    c) Both of these
    d) None of the above

47. Rate at which the ramp angle changes in degrees per 100ft which is measured in S-shape holes is called
    a) Drop rate
    b) Build rate
    c) Both of these
    d) None of the above

48. The connection between joints of the drill pipe is called
   a) Tool joint
   b) Pipe joint
   c) Blast joint
   d) None of the above

49. When are drilling jars used?
   a) To free a stuck drill pipe
   b) To stop the mud circulation
   c) To prevent lost circulation
   d) All of the above

50. How are drilling motors powered?
   a) By flow of mud through the motor
   b) By the use of electricity
   c) Both of these
   d) None of the above

51. Factors influencing bit selection include
   a) Formation lithology
   b) Rig capabilities
   c) Economic goals
   d) All of the above

52. The radii of short radius horizontal wells are
   a) 2-6 ft
   b) 300-900 ft
   c) 1000-3000 ft
   d) None of the above

53. The radii of medium radius horizontal wells are
   a) 2-6 ft
   b) 300-900 ft
   c) 1000-3000ft
   d) None of the above

54. The radii of long radius horizontal wells are
   a) 2-6 ft
   b) 300-900 ft
   c) 1000-3000
   d) None of the above

55. Which of the following factors is essential to hole cleaning?
   a) Flow rate
   b) Rheology
   c) Pipe Rotation
   d) Drilling rate

56. The storage device for nitrogen pressurized hydraulic fluid which is used in operating blowout preventers
   a) Accumulator
   b) Lubricator
   c) Riser
   d) None of the above

57. The depth at which the well is deflected from the vertical is known as
   a) Kick off point
   b) True vertical depth
   c) Both of these
   d) None of the above

58. The trajectory of a directionally drilled well in three dimensions is called
   a) Well path
   b) Well trajectory
   c) Well Locus
   d) None of the above

59. The length of the wellbore measured along the actual well path is known as

a) Measured depth
b) True vertical depth
c) Both of these
d) None of the above

60. The vertical distance from a point in the well (current or final depth) to the surface (rotary kelly bushing) is known as

a) True vertical depth (TVD)
b) Total depth
c) Casing depth
d) None of the above

61. "Journal Angle" is related to-

a) Rig design
b) Casing design
c) Drill string design
d) Bit design

62. A measure of how easy the formation is to drill is known as

a) Drillability
b) Abrasiveness
c) Both of these
d) None of the above

63. A measure of how rapidly the teeth of a milled tooth will wear when drilling the formation is known as

a) Abrasiveness
b) Corrosiveness
c) Journal angle
d) Fleet angle

64. Which of the following is correct?

a) The abrasiveness tends to increase as the drillability decreases
b) The abrasiveness tends to increase as the drillability increases
c) Abrasiveness doesn't affect drillability
d) None of the above

65. The main objectives of well planning include

a) Safety
b) Usable hole
c) Minimize cost
d) All of the above

66. The process of directing the well bore along some trajectory to a predetermined target is called

a) Directional drilling
b) Side tracking
c) Horizontal drilling
d) None of the above

67. Deviation of the well around and away from an obstruction in the original well bore such as a stuck pipe is called

a) Sidetracking
b) Horizontal drilling
c) Multilateral drilling
d) None of the above

68. The angle measured in degrees between the actual well path at some depth and a vertical line directly below the rig site is called

a) Drift or inclination angle
b) Fleet angle
c) Azimuth
d) Journal angle

69. Types of well trajectories include
    a) build and hold
    b) S-type - bulid->hold->drop
    c) Modified S-type - build->hold->drop- >hold
    d) All of the above

70. BOP consists of:
    a) Pipe Ram
    b) Shear Ram
    c) Blind Ram
    d) All of the above

71. Which of the following is false regarding the pipe ram?
    a) It should have semi-circular openings
    b) It must match the size of pipe currently in use
    c) It cannot be used when the drillstring is out of the hole
    d) It can close of a tool-joint

72. Which type of BOP is capable of closing on any piece of equipment?
    a) Pipe ram
    b) Blind ram
    c) Shear ram
    d) Annular preventer

73. Which of the following statements is wrong regarding the mud telemetry system:
    a) It transmits mud pulses through the mud column
    b) It is used in both compressible and incompressible fluids
    c) It requires clean mud circulation without fiber
    d) It's data transmission rate is limited

74. Measurement while drilling (MWD) systems provide real-time:
    a) Drillbit information
    b) Borehole inclination
    c) Azimuth
    d) All of the above

75. Which of the following statements is false about standard fluid loss test?
    a) Test duration is 15 minutes
    b) The test is conducted at 100 psi
    c) B and C
    d) None of the above

76. Which of the following shows the correct order of solid removal equipment used when unweighted mud is reconditioned?
    a) Centrifuge > shale shaker > desander > desilter
    b) Desilter > centrifuge > shale shaker >desander
    c) Shale shaker > desander > desilter >centrifuge
    d) Desander > desilter > centrifuge > shale shaker

77. Hydrocyclones cannot be used alone with weighted mud systems because:
    a) Their cut off point falls in the particle-size range of barite

b) They are very expensive

c) They are not effective in removing coarse particles

d) Their operating cost is very high

78. The tendency of differential pressure sticking is high when:

a) Overbalance pressure is low

b) Mud cake thickness is low

c) Both overbalance pressure and mud cake thickness are high

d) Annular diameter ratio is low

79. Which of the following statements is incorrect about blind rams?

a) They are designed to close when no pipe is in the hole

b) They flatten drill pipe if accidentally used

c) They can stop the flow from the well if used when the drill string is in the hole

d) (a) and (c)

80. The tension in the fast line must be kept:

a) Below the nominal breaking strength

b) Above the nominal breaking strength

c) Exactly at the nominal breaking strength

d) All of the Above

81. Which of the following is not a function of well control system?

a) Detecting a kick

b) Allowing mud and kick fluid circulation under pressure

c) Allowing pipe movement under pressure

d) None of the Above

82. Which of the following statements is false about pipe ram?

a) It has semicircular openings, which match the diameter of the sizes for which they are designed

b) It size must match the size of pipe currently in use

c) If more that one size of drill pipe is in the hole, additional ram preventers must be used in the BOP stack

d) They can close on a Kelly

83. Which of the following is the part of BOP?

a) Flow line

b) Kill line

c) Wire line

d) Drill line

84. Which of the following lifting operation is related to fishing job?

a) Broken well head

b) Broken Christmas tree

c) Broken drill bit

d) Broken storage tank

85. Which of the following shows the correct mud circulation path during conventional drilling?

a) Suction tank ->drill-string->mud pump- >annular space->Shale shaker

b) Suction tank->shale shaker->mud pump->drillstring->drill bit->annular space

c) Suction tank->mud pump->drillstring->drillbit->annular space->shale shaker

d) Suction tank->drill bit->mud pump->drillstring->annular space->shale shaker

86. All tubulars are specified by:

a) Range, weight per foot, outside diameter, steel composition

b) Range, weight per foot, inner diameter, steel grade

c) Manufacturers name, weight per foot, outside diameter, steel grade

d) Range, Weight per foot, outside diameter, steel grade

87. Surge chamber is used after the mud pump to:

a) Dampen pressure surges and flow pulsations

b) Remove solid particles

c) Prevent barite settling

d) Stabilize the draw works

88. A kick has been received while the bit was off the bottom. Which preventer can be used to lower the bit to the bottom of the hole to perform kill operation?

a) Pipe Ram

b) Shear Ram

c) Annular Preventer

d) Blind Ram

89. The depth of the well governs:

a) The lithology that must be penetrated

b) The time required to complete the well

c) (a) and (b)

d) None of the above

90. If the formation pressure is greater than the normal pressure at a given depth, then it is called:

a) Normal pressure

b) Abnormal pressure

c) Subnormal pressure

d) None of the above

91. Which of the following is not the objective of drilling operations?

a) Minimize the total well cost

b) Drill a useable hole

c) Minimize non-productive rig downtime

d) Drill well in a safe and environmentally sound manner

92. Which of the following fluid is recommended to drill reactive shale zones?

a) Water-based mud

b) Oil-based mud

c) Both (a) and (b)

d) None of the above

93. Duplex pumps:

a) Are non-reciprocating type

b) Are single-acting triplex

c) Have constant horse power

d) None of the above

94. The trip time primarily depends on:
   a) Depth of the well
   b) Rig type and capabilities
   c) Drilling method
   d) All of the above

95. Which of the following bits are recommended for drilling hard formations in the deep section of the well?
   a) Roller bit with tungsten carbide insert
   b) Roller bit with mill tooth
   c) PDC bit
   d) Diamond bit

96. Which of the following is false about diamond bits?
   a) They set in a tungsten carbide matrix
   b) Under proper bit operation only the diamond contact the hole bottom
   c) Size and number of diamond used depends on the hardness of the formation to be drilled
   d) Bit for hard formations should have many large stones

97. Which of the following statements is false regarding the shape of a diamond bit?
   a) Long taper diamond bits are used to drill straight hole with higher bit weight
   b) Short taper bits are difficult to clean
   c) A more concave bit design is used for directional drilling application
   d) All of the above

98. Rock cutting mechanisms of drag bits are primarily by:
   a) Scraping and grinding actions
   b) Crushing action
   c) Wedging action
   d) Twisting action

99. Roller cone bits works by
   a) Shearing or cutting the formation.
   b) crushing the formation
   c) both (a) and (b)
   d) None of the above

100. Which one of the following is false regarding cuttings transport ratio?
   a) For good hole cleaning, the cuttings transport ratio must be more than unity
   b) When the particle slip-velocity of zero, the cutting transport ratio is unity
   c) As the slip velocity increases, the transport ratio increases
   d) As the slip velocity decreases the concentration of cuttings in the annulus decreases

101. Which of the following is a major consideration for cutting transport in horizontal wells?
   a) Annular size
   b) Drilling rate
   c) Eccentricity
   d) Rotary speed

102. Following dial readings have been obtained from Fann V-G meter for a drilling mud.

Reading at 600 rpm = 60

Reading at 300 rpm = 40

What are the values of apparent viscosity and plastic viscosity?

a) 30, 20
b) 30,10
c) 30,30
d) None of the above

103. Conditions necessary for the development of abnormal pressure in the formation are:

a) Mechanism causing formation pressure to change
b) The presence of permeable formation layer
c) The presence of impermeable barrier
d) (a) and (c)

104. Conventional (overbalanced) drilling means:

a) The borehole pressure is greater than the formation pressure
b) The borehole pressure is lower than the formation pressure
c) At the surface, the annulus is closed during normal drilling operation
d) All of the above

105. Which one of the following is correct about drilling cost?

a) It doesn't vary with well depth
b) It decreases exponentially with well depth
c) It is independent of the well depth
d) It increases exponentially with well depth

106. Drilling cost depends primarily on what?

a) Well location and depth
b) Depth
c) Settling, Suction and Reserve
d) (a) &(b)

107. Filter cake thickness is measured and reported to

a) 1/32 of an inch
b) 1/64 of an inch
c) 1/16 of an inch
d) 1/48 of an inch

108. Viscosity of a drilling fluid (Newtonian fluid) is determined using Fann V-G viscometer. The dial reading at a rpm of 600 is 60, then apparent viscosity of the fluid in cP is

a) 30
b) 90
c) 60
d) 50

109. A mud sample in a rotational viscometer equipped with a standard torsion spring gives a dial reading of 46 when operated at 600 rpm and a dial reading of 28 when operated at 300 rpm. The yield point for this mud will be -----lb/100 ft2

a) 10
b) 5
c) 15
d) 18

110. The mud density resulting from the addition of 100 barrels of 42 °API oil to 800 barrels of 11.3 ppg mud will be -------

a) 12.5 ppg

b) 9. 3 ppg

c) 10.8 ppg

d) 8.5 ppg

111. A low pressure zone at 700 feet will support a 7.5 lb/gal column of mud. If 8.33 lb/gal water and air are mixed. Assume a pump rate of 420 gpm. The volume of air requirement will be...

a) 0.2545 cubic ft air/cubic ft mud

b) 2.545 cubic ft air/cubic ft mud

c) 1.254 cubic ft air/cubic ft mud

d) 0.1254 cubic ft air/cubic ft mud

112. The A well is drilled at 8000 ft with 11.0 ppg mud and surface casing is set at 2500 ft. The hydrostatic pressure at surface casing is ---

a) 2040 psi

b) 1000 psi

c) 1430 psi

d) 1650 psi

113. What is the bottom hole hydrostatic pressure of a 12.1 pound mud at 10, 500'?

a) 6606.6 psi

b) 6500 psi

c) 7500.5 psi

d) 8650 psi

114. Which of the followings fluid requires a certain level of pressure gradient to induce flow in a pipe?

a) power law fluid

b) newtonian fluid

c) bingham plastic fluid

d) all of the above

115. Which of the following is not the function of drilling fluid?

a) transport rock cuttings to the surface

b) increase the reservoir pressure

c) control the formation pressure and prevent walls from caving

d) transmit hydraulic power to drill motor, tools, and bit

116. Marsh funnel viscometer is used to measure:

a) Fluid loss

b) Cake thickness

c) Viscosity

d) None of the above

117. Viscous properties of oil-based mud is primarily controlled by

a) Viscosifier content

b) Oil-water ratio

c) Barite content

d) None of the above

118. Mud density is measured using

a) Rotational viscometer

b) Marsh funnel

c) Mud balance

d) Filter press

119. Which of the following statements is not true about drilling muds/fluids?
   a) Drilling muds are complex fluids.
   b) Drilling fluids can contain a number of additives and contaminants.
   c) Drilling fluid related problems strongly affect the number of rig days
   d) Choice and control of drilling fluids are not the concern of all drilling personnel.

120. The advantages of air and mist drilling fluids include ______
   a) Higher penetration rates
   b) Longer bit life
   c) No lost circulation problems
   d) All of the above

121. Which of the following are the disadvantage of air is mist drilling fluids?
   a) Control of fluid influx
   b) Control of high-pressure zones
   c) Both (a) and (b)
   d) None of the above

122. Which of the following is true about the advantages of air and mist drilling fluids over conventional drilling fluids?
   a) Control of fluid influx
   b) Higher penetration rates
   c) Control of high-pressure zones
   d) None of the above

123. Which of the following fluid is preferred to drill low pressure depleted zones?
   a) Water-based mud
   b) Oil based mud
   c) foam
   a) All

124. Which of the following causes the montmorillonite structure to have an excess of electron?
   a) Contamination of montmorillonite with illite
   b) Mixing of montmorillonite with chlorite
   c) Mixing of montmorillonite with kaolinite
   d) Replacement of some of the aluminum ions by magnesium ions in the crystal structure

125. Which of the following property measures the cutting suspension capacity of drilling fluid?
   a) Apparent viscosity
   b) Plastic Viscosity
   c) Gel strength
   d) None of the above

126. Which of the following additive is used to reduce the density of the cement slurry and to increase slurry yield?
   a) Bentonite
   b) pozzolan,
   c) gilsonite
   d) All of the above

127. Which of the following is true about double-acting duplex pumps?

a) It has three pistons

b) It sucks and discharges on every stroke

c) The displaced volume is only a function of linear diameter

d) The hydraulic power is a function of the pressure differential between the input and out and fluid temperature

128. Kick volume is normally determined from:

a) Shut in drill pipe pressure

b) Density of the kick

c) Pit gain

d) Shut in casing Pressure

129. Which of the following statement is false about different size nozzles installed on a drill bit?

a) the pressure drops across the nozzle are the same

b) the flow rates of the nozzles are the same

c) the nozzle velocities are the same

d) None of the above

130. Which of the following is not optimization parameters used for selecting bit nozzle size?

a) Bit Nozzle Velocity

b) Bit hydraulics horse power

c) Turbulence occurs when the flow becomes stable

d) All of the above

131. Which of the following statement is true about turbulence?

a) turbulence decreases pressure loss

b) Turbulence increases pressure loss

c) Turbulence occurs when the flow becomes stable

d) All of the above

132. Where is production casing normally set?

a) Below the production zone

b) Above conductor casing

c) Below the conductor casing

d) None of the above

133. Which of the following factors has the least impact on casing seat selection?

a) Rate of Penetration

b) Pore pressure

c) Kick tolerance

d) Fracture gradient

134. When considering calculation methods for survey points, the least precise method is:

a) Circular arc

b) Minimum curvature

c) Radius of curvature

d) Tangential

135. Which of the following is not a typical purpose for a BHA?

a) Assure casing can be run into a hole

b) Control direction and inclination in directional holes

c) Increase severity of doglegs, keyseats and ledges

d) Protect the drill pipe from excessive bending and torsional loads

136. Which of the following is a typical characteristic of plug flow cement placement?

a) Reynolds number less than 100

b) Large cement volumes

c) High displacement rates

d) Short cementation times

137. Which of the following is not a typical cause for a fishing job?

a) Cementing error

b) Differential sticking

c) Swelling zone

d) Use of oil-base mud

138. Which of the following is not a component of the hoisting equipment for a drilling rig?

a) Crown block

b) Drawworks

c) Drilling line

d) Rotary table

139. Which of the following is not a primary function of drilling fluids?

a) Cuttings removal

b) Frac proppant

c) Lubrication and cooling

d) Well control

140. Which of the following methods is most likely to hydrostatically balance a well kick with only one circulation?

a) Concurrent

b) Driller's

c) Toolpusher's

d) Wait and weight

141. Which of the following is not a colloidal polymer?

a) Xanthan gum

b) Starch

c) Sodium carboxy methyl cellulose

d) Hydroxyethyl cellulose

142. Which of the following is a major consideration for cuttings transport in horizontal wells?

a) Annulus size

b) Drilling Rate

c) Eccentricity

d) Rotary speed

143. In comparing the API specs for L-80 and N-80 tubulars, which of the following statements is true?

a) Minimum yield strength is different

b) Maximum yield strength is the same

c) Minimum tensile strength is different

d) Minimum elongation is the same

144. Which of the following is not a typical casing hanger type?

a) Slip-weld

b) Nominal flange

c) Boll-weevil

d) Automatic

145. The control pressure for a surface safety valve is most likely equal to:

a) Valve body pressure divided by actuator ratio

b) Actuator ratio divided by valve body pressure

c) Two times valve body pressure divided by actuator ratio

d) Valve body pressure multiplied by actuator ratio

146. Which of the following has the least influence on drill pipe stretch?

a) BHA weight

b) Drill pipe weight

c) Mud density

d) Traveling block weight

147. Which of the following is most likely a descriptor for a diamond bit?

a) Crown

b) Journal

c) O-ring Seal

d) Shirttail

148. Which of the following is not a description of fluid flow?

a) Laminar

b) Plastic

c) Plug

d) Turbulent

149. Which of the following is not a key rheological property of drilling fluids?

a) Yield point

b) Plastic viscosity

c) Gel Strength

d) Funnel viscosity

150. Which of the following has the least influence on bit nozzle pressure drop?

a) Bit diameter

b) Flow rate

c) Mud density

d) Nozzle area

151. Geo-steering is a tool to control

a) Formation pressure

b) Rheology of drilling fluid

c) Drilling trajectory

d) None of the above

152. Which of the following is not a typical section view for directional wells?

a) Double build

b) Double twist

c) S type

d) Slant type

153. Which of the following is a advantage of air drilling?

a) Differential pipe sticking is eliminated

b) Downhole explosions or fire may occur

c) Lost circulation may be reduced

d) Reduced thawing of frozen zones

154. Which of the following is false regarding leak-off tests?

a) Unpurged air can affect the shape of the leak-off test curve

b) There are two phases of fracturing a formation, initiation and extension

c) The pressure to initiate is usually less than the pressure to extend

d) The initial pressure increase vs. volume pumped reflects elastic rock deformation

155. Which of the following is most likely regarding cement flow regimes?

a) Laminar flow has the smallest displacement efficiency

b) Turbulent flow gives low displacement efficiency

c) Turbulent flow requires the longest pumping times

d) None of the above

156. When backing-off a stuck drill string, which of the following would not be a typical step?

a) Twist the string one counter-clockwise turn per 1,000 feet of free string

b) Pull the buoyed weight of the string above the stuck point

c) Fire the shot after working the torque down to the tool joint to be backed off

d) Twist the string one clockwise turn per 1,000' of free string

157. Suggested guideline for the steps required in the well profile design:

a) Kick off point determination.

b) Build up rate and tangent section

c) TVD uncertainty of reservoir and geological markers

d) All of the above

158. Which of the following is not a typical contributing factor in axial friction forces?

a) Bending of the drill string in a dog leg

b) Buoyed weight of the drill string lying on the low side

c) Hanging up of drill string component on a ledge

d) Packing of the drill string in a cuttings bed

159. Which of the following does not result in collapse of the tube?

a) Elastic Collapse Pressure

b) Plastic Collapse Pressure

c) Transition Collapse Pressure

d) Yield-Strength Collapse Pressure

160. The buoyancy factor of 1.3 gm/cc mud will be

a) 0.834

b) 0.734

c) 0.834

d) None of the above

161. If the maximum weight on bit is anticipated to be 20 tonne, the average drill collar length is 8.2 m, the drill collar weight is 0.320 tonne/m and buoyancy factor of the mud is 0.847, The number of drill collars required will be:

a) 9

b) 10

c) 11

d) 12

162. What will be actual load on derrick if the suspended load weight 40 tonnes and the block is strung up on 10 lines?

a) 45

b) 48

c) 53

d) 55

163. A caliper survey has determined that there has been a uniform 65% wall loss in a 5-in. (127 mm) 23.2 lbm/ft (34.5 kg/m) T95 casing string. Neglecting axial/hoop stresses. What is the new burst pressure rating of this casing string in psi/(Mpa)?

a) 5,550 (38.3)

b) 6,350 (43.78)

c) 10,340 (71.29)

d) 14,300 (98.6)

Statement for Linked Answer Questions 164-166

A drill string with a buoyant weight of 200000 lbs must be pulled from the well. A total of 8 lines are strung between the crown block and the travelling block. Assume that a four sheave, roller bearing system is being used.

164. The tension in the fast line will be :

a) 29691 lbs

b) 19691lbs

c) 39591lbs

d) 49691lbs

165. The tension in the dead line will be :

a) 25000 lbs

b) 50000 lbs

c) 35000 lbs

d) 60000 lbs

166. the vertical load on the rig when pulling will be:

a) 154691 lbs

b) 254691 lbs

c) 354691 lbs

d) 454691 lbs

167. A well is drilling ahead at 10,000 ft MD (9,000 ft TVD) with a BHA comprised of 400 ft of 8-in × 3-in drill collars (147 lb/ft) and 4 1/2-in, 16.60 lb/ft drill pipe (18.40 lb/ft adjusted weight). A bit trip is required. Assume there is no fluid lost to the formation and the drill string is pulled dry (i.e., there is no loss of fluid at the surface). The total volume (barrels) of drilling fluid required to keep the hole full while tripping completely out of the hole is most nearly:

a) 54

b) 67

c) 79

d) 86

168. A production string of 7-inch P-110 casing with a nominal wall thickness of 0.590 inch and a nominal weight per foot of 41 lb/ft is run to TD. Assume yield strength mode of failure in collapse given the D/t range that this particular pipe falls into and that the pipe is submitted

to stresses of 50,000 psi axial tension and 11,000-psi internal pressure. With a safety factor of 1.0, the minimum pressure (psi) that would cause collapse in these conditions is most nearly:

a) 27,980

b) 26,684

c) 21,186

d) 16,980

169. A vertical well is drilling ahead at 10,000 ft MD. A bit trip is required and the results of the hydraulics optimization suggest a flow rate of 500 gpm with three 13/32 nozzles (nozzle area = 0.3889 in2). The drilling fluid has a density of 10.0 lb/gal and behaves like a Bingham plastic fluid with a PV of 40 cp and YP of 15 lbf/100 ft2. Assume the internal pipe velocity will be 13.953 ft/min and is in the turbulent flow regime with the pressure drops in the annulus and the surface equipment considered negligible. Assuming 100% mechanical efficiency, the required surface horsepower from the mud pumps to satisfy the proposed hydraulics optimization is most nearly:

a) 876

b) 1,479

c) 1,523

d) 3,002

170. While drilling ahead at 10,000 ft MD, a twist-off occurs in the body of the drill pipe. The portion of the body recovered indicates that the top of the fish has a lip that is curled inward. The option that has the most likely chance of recovering the fish is most nearly:

a) Basket overshot

b) Bulldog spear

c) Rope spear

d) Wash pipe

171. Fracture gradients need to be predicted to support a casing program design for a land-based operation in a sedimentary basin. Assume that two primary methods are available for predicting fracture gradient. Method 1 is the Matthew and Kelly model and Method 2 is the Eaton model.

The statement that is false is:

a) Method 1 assumes a constant regional overburden stress gradient, and Method 2 assumes a variable overburden stress gradient.

b) Both methods accommodate formation pressures that are greater than a normal pressure gradient.

c) Both methods accommodate formation pressures that are greater than the fracture pressure gradient.

d) Both methods assume the mechanical strength of the rock increases with depth.

172. A 6×18-inch (liner size × stroke length) duplex double acting pump has a volumetric efficiency of 87%. The rod size is the standard 2 inch. The pump rate is 48 spm. The 10,000-ft, $4^1/2$-inch annulus (10000ft) has a capacity of 0.05 bbl/ft. The vol-

umetric output per minute will be

a) 8.25 bbl/min.

b) 10.5 bbl/min

c) 12.0 bbl/min

d) 7.25 bbl/min

173. A directional well is to be drilled with a build and hold profile due north with a proposed kick-off point of 3,000 ft MD (3,000 ft TVD), a constant assumed build rate of 3°/100 ft, and a proposed target of 8,000 ft TVD with a horizontal departure of 1,750 ft. The measured depth (ft) at the end of the build section is most nearly:

a) 1,910

b) 3,687

c) 4,624

d) 8,311

174. A well is drilling ahead with a 9.0-lb/gal brine. The maximum desired particle size is 20 microns. The pit setup is conventional with the flow line discharging into the possum belly, and the centrifuge overflow will be discharged to the reserve pit, if that equipment is selected. The shaker is equipped with 20 mesh screens. The daily rental rates for solids control equipment are shaker, US$20; desander, US$25; desilter, US$30; and centrifuge, US$50. The equipment combination that gives the **least** daily cost while satisfying the given design and operational guidelines is most nearly:

a) Shaker

b) Shaker; centrifuge

c) Shaker; desander; desilter

d) Shaker; desander; desilter; centrifuge

175. A fluid-filled, vertical well where the MD equals the TVD is drilling ahead with a BHA of fixed length and with a drilling fluid density greater than 8.3 lb/gal. The appropriate available bit weight needs to be identified to minimize drilling costs while maintaining drill string integrity. Two primary methods are available to calculate the maximum available weight on bit while remaining within design constraints. Method 1 designs on buoyancy and places the neutral point of buckling at the top of the BHA for maximum available weight on bit. Method 2 designs on pressure-area summation of forces and places the axial forces equal to zero at the top of the BHA for maximum available weight on bit. The MD is greater than the fixed length of the BHA. The statement that is true regarding the maximum available weight on bit for the same fixed BHA length is:

a) Method 1 will yield a higher maximum available weight on bit than Method 2.

b) Method 2 will yield a higher maximum available weight on bit than Method 1.

c) Both methods will result in the same maximum available weight on bit.

d) The given information is inadequate to make a prediction between the two methods.

176. State regulations require a cement plug from the bottom perf to 100 ft above the top perf. You plan to lay a single balanced cement plug that will fulfill the requirements to the letter. Assume no excess is required and the weights of the fluid in the hole and the displacing fluid are equal. The following information is given:

Casing 7-inch, 23-lb/ft cemented at 4,040 ft

Plug back depth 4,000 ft

Perforations 3,900 ft−3,700 ft and 3,650 ft−3,600 ft, four 1/2-inch JHPF

Work string 3½-inch, 9.3-lb/ft EUE tubing

Excluding pumps and lines, the volume (bbl) of displacement required is most nearly:

a) 31.6

b) 31.3

c) 31.1

d) 30.1

177. A well has a proposed kick-off point of 12,000 ft MD (12,000 ft TVD) with 0 degrees of deviation and with a final proposed total depth of 13,500 ft MD (12,100 ft TVD) at 90 degrees of deviation. The 90 degrees of deviation was initially achieved at 12,157 ft MD. The directional drilling technique that most nearly addresses the desired wellbore trajectory is:

a) Short radius

b) Medium radius

c) Long radius

d) Extended reach

178. A well was drilling ahead and experienced a fishing job. The daily cost of continued fishing operations is US$11,500. The total cost of the fish, if left in the hole, is US$165,000. Continued fishing operations have a 30% probability of success. The cost of sidetracking is estimated to be US$250,000 with an assumed 100% probability of success. To the nearest whole day, when is it more expensive to continue fishing operations on a risked dollar basis versus the cost of sidetracking?

a) 8

b) 11

c) 15

d) 22

179. A well is drilling ahead and experiences a fishing job that results in the need to set a plug for sidetracking. A 2 3/8-inch fiberglass stinger is run in conjunction with the 5-inch drill pipe. A balanced plug is set and the hole is not kept full while pulling out of the cement plug with the stinger to minimize fluid disturbance. The statement that most nearly describes a drawback in this balanced plug method is:

a) The drill pipe will typically pull dry.

b) The fiberglass stinger allows the plug to remain drillable if the stinger becomes stuck.

c) Contamination of the cement slurry with mud is minimized.

d) The internal capacity and displacement volume difference between the drill pipe and the stinger is not accounted for.

180. A well is drilling ahead and enters a gas-bearing, permeable formation. A gas influx of a known size enters into the wellbore. Assume the influx is contiguous (no mixing with the drilling fluid) and no gas migration occurs. The measured depth equals the true vertical depth, the O.D. of the drill string is constant including the BHA, and the open hole O.D. and casing I.D. are equal. The volume of the influx is less than the annular volume between the open hole and drill string. The influx is circulated out using a constant, balanced bottomhole pressure method that allows no additional influx while circulating only original density drilling fluid. The statement that is most correct in describing the expected maximum pressure at the casing shoe is:

a) The maximum pressure occurs as the gas reaches the surface when the surface casing pressure is maximum.

b) The maximum pressure occurs when the gas bubble/ drilling fluid Interface is at the casing shoe.

c) The maximum pressure occurs when the gas column is at its minimum height during the initial shut-in of the well. When the maximum pressure occurs is not predictable.

d) When the maximum pressure occurs is not predictable.

181. A well is drilling ahead in a fluid-filled vertical hole where MD equals TVD and the drilling fluid density is >8.3 lbm/gal. The next bit run requires optimum hydraulics to minimize drilling costs. Two primary methods are available to optimize the bit hydraulics. Method 1 will optimize bit hydraulic horsepower, and Method 2 will optimize jet impact force. Assume that the surface pressure is the same for both methods, that adequate surface horsepower is available from the mud pumps for both methods, and that the optimum flow rate for both methods lies between the minimum acceptable annular velocity and the maximum acceptable annular velocity. The statement that is most correct regarding the optimum flow rate is:

a) The optimum flow rate for both methods is the same since the surface pressure is the same.

b) The optimum flow rate for Method 1 will be greater than the optimum

c) The optimum flow rate will be greater for Method 1 50% of the time, and the remaining times will be greater for Method 2.

d) The optimum flow rate for Method 2 will be greater than the optimum flow rate for Method.

Statement for Linked Answer Questions 182-183

A 9.5 lb/gal drilling mud contains clay (Sp. Gravity = 2.5) and fresh water for the drilling of an oil well.

182. The volume % of the solids is:
   a) 10.5%
   b) 9.4%
   c) 12.6%
   d) 13.5%

183. The weight % clay in this mud is:
   a) 20.6%
   b) 22.5%
   c) 28.5%
   d) 15.2%

Statement for Linked Answer Questions 184-185

For laboratory purposes, it is desired to mix one liter of bentonite-fresh water mud having a viscosity of 30 cp and solid content of 3.0% by volume.

184. What will be the resulting mud density?
   a) 1.045 gm/cc
   b) 1.248 gm/cc
   c) 1.349 gm/cc
   d) 1.545 gm/cc

185. How much of each material should be used?
   a) 75 gm bentonite and 1000 cc water
   b) 75 gm bentonite and 970 cc water
   c) 75 gm bentonite and 975 cc water
   d) 75 gm bentonite and 925 cc water

186. How much weighting material ($BaSO_4$, the mineral barite, sp. Gr. = 4.3) should be added to the mud (density 8.7 lb/gal) to increase its density to 10 lb/gal and what will be the resulting volume?
   a) 217 gm and 1050 cc
   b) 217 gm and 983 cc
   c) 217 gm and 1000 cc
   d) 217 gm and 783 cc

187. How many sacks of barite are necessary to increase the density of 1000 bbl of mud from 10 to 14lb/gal and what will be the final mud volume?
   a) 2750 and 1250 bbl
   b) 2750 and 1180 bbl
   c) 2650 and 1250 bbl
   d) 2650 and 1180 bbl

188. How much fresh water must be added to 1000 bbl of 12 lb/gal mud to reduce its density to 10 lb/gal and what will be the resulting volume?
   a) 1200 bbl and 2100 bbl
   b) 1000 bbl and 2000 bbl
   c) 1100 bbl and 2100 bbl
   d) 1200 bbl and 2200 bbl

189. A mud engineer finds from pilot tests that 2.0 gm of CMC is required to obtain the desired water loss reduction for a one liter mud sample. How much CMC should be added to the actual 1000 barrel system?
   a) 650 lb
   b) 500 lb
   c) 700 lb
   d) 350 lb

190. What would be the cycle time for the conditions given below?

V_m = 1000 bbl

Pump liners = $7^1/2$ inch diameter

Stroke length = 16 inch

Piston rod diameter = 2 inch

N = 40 strokes per minute

Power pump is used

a) 120 min

b) 150 min

c) 100 min

d) 110 min

191. A string of inch, 13.3 lb/ft grade E drill pipe is stuck in a 10,000 ft hole. The driller obtains the following data as described previously?

F_1 = 140, 000 lb, which is greater than the weight of string.

F_2 = 200,000 lb, which is less than the yield strength of the pipe

e = 4 ft

What is depth to stuck pipe point?

a) 5600 ft

b) 7800 ft

c) 8200 ft

d) 6800 ft

192. Casing is being cemented in a 10,000 ft well containing 10 lb/gal mud. The slurry density is 14 lb/gal and the plug is to be chased with 0.85 sp. gr. oil. If the anticipated fill-up is 1000ft, what surface pressure must be held on the casing to prevent backflow at the end of the job in case the floating equipment fails to hold?

a) 1200 psi (surface pressure)

b) 1900 psi (surface pressure)

c) 13500 psi (surface pressure)

d) 1700 psi (surface pressure)

193. A 10 lb/gal mud is being circulated at the rate of 500 gal/min through a tri-cone bit having three inch diameter jets. What is the pressure drop across the bit? Given: C = 0.95.

a) 2100 psi

b) 2500 psi

c) 2800 psi

d) 1500 psi

194. Given data : Annular pressure loss = 200psi

True vertical depth = 10,000ft

Mud weight = 9.6ppg

The Equivalent Circulation Density of Mud will be:

a) 12.5 ppg

b) 11.5 ppg

c) 10.0 ppg

d) 12.0 ppg

195. Given data: Leak-off test pressure = 1140psi

Casing shoe tvd = 4000ft

Mud weight = 10.0ppg

The Maximum allowable mud weight (ppg) will be:

a) 17.50 ppg

b) 15.48 ppg

c) 12.52 ppg

d) 11.50 ppg

196. The pump output, gpm, for a 7-in. by 12-in. triplex pump at 80 strokes per minute will be:

a) 655.5gpm

b) 555.5gpm

c) 355.5gpm

d) 455.5gpm

197. Given data: bbl/stk, of a 5-1/2-in. by 14-in. duplex pump @ 100% efficiency. Rod diameter = 2.0in.

The pump output in bbl/stk will be

a) 0.30892

b) 0.10892

c) 0.40892

d) 0.60892

198. What will be the hydrostatic pressure decrease when pulling dry pipe out of the hole for the following given data?

Given data:

Number of stands pulled = 5

Average length per stand = 92ft

Pipe displacement = 0.0075 bbl/ft

Casing capacity = 0.0773 bbl/ft

Mud weight = 11.5ppg

a) 18.5 psi

b) 21.5 psi

c) 29.5 psi

d) 35.5 psi

199. What will be the hydrostatic pressure decrease when pulling wet pipe out of the hole for the given data?

Given data:

Number of stands pulled = 5

Average length per stand = 92ft

Pipe displacement = 0.0075 bbl/ft

Pipe capacity = 0.01776 bbl/ft

Casing capacity = 0.0773 bbl/ft

Mud weight = 11.5ppg

a) 133.5 psi

b) 152.5 psi

c) 165.5 psi

d) 173.5 psi

200. What should be the feet of dry pipe that must be pulled to lose the overbalance using the given data?

Given data:

Amount of overbalance = 150 psi

Casing capacity = 0.0773 bbl/ft

Pipe displacement = 0.0075 bbl/ft

Mud weight = 11.5ppg

a) 2334

b) 2800

c) 2659

d) 2925

201. What should be the feet of wet pipe that must be pulled to lose the overbalance using the given data?

Given data:

Amount of overbalance = 150 psi

Casing capacity = 0.0773 bbl/ft

Pipe capacity = 0.01776bbl/ft

Pipe displacement = 0.0075 bbl/ft

Mud weight = 11.5ppg

a) 582.6

b) 542.8

c) 550.5

d) 516.8

202. What should be the new circulating pressure, psi of a pump using the given data?

Given data

Present circulating pressure = 1800 psi

Old pump rate = 60 spm

New pump rate = 30 spm

a) 450 psi
b) 550 psi
c) 650 psi
d) 400 psi

203. Given data:

Desired length of dry pipe (2 stands) = 184ft

Mud weight = 12.2ppg

Slug weight = 13.2ppg

Drill pipe capacity = 0.01422bbl/ft

The barrels of slug required for the above data will be:

a) 62.0
b) 42.0
c) 32.0
d) 52.0

204. Given Data:

Desired WOB while drilling = 50,000 lb

Safety factor = 15.0%

Drill collar weight = 147 lb/ft

Mud weight = 12.0 ppg

8 inch OD- 3.0 inch ID

The length of bottomhole assembly necessary for a desired weight on bit will be:

a) 499 ft
b) 479 ft
c) 469 ft
d) 429 ft

Statement for Linked Answer Questions 205-206

Given data:

Class H cement plus 6% bentonite to be mixed at 14.0 lb/gal

Specific gravity of bentonite = 2.65.

205. The weight of materials, lb/sk will be:

a) 166.3
b) 150.5
c) 175.5
d) 189.5

206. The volume of slurry, gal/sk will be :

a) 22.56
b) 15.76
c) 11.86
d) 19.26

207. Water requirement using material balance equation will be:

a) 8.0 gal/sk of cement
b) 5.0 gal/sk of cement
c) 10.0 gal/sk of cement
d) 12.0 gal/sk of cement

208. What would be the amount of hematite, lb/sk of cement e required to increase the density of Class H cement to 17.51b/gal?

Given: Water requirement of cement = 4.3 gal/sk

Water requirement of additive (hematite) = 0.34gal/sk

Specific gravity of cement = 3.14

Specific gravity of additive (hematite) = 5.02

a) 18.1 lb of hematite per sk of cement

b) 19.1 lb of hematite per sk of cement

c) 15.1 lb of hematite per sk of cement

d) 12.1 lb of hematite per sk of cement

209. Given data:

Mud weight = 12.5 ppg

Weight of water = 8.33 ppg

TVD = 10,000 ft

Annular capacity = 0.1279bbl/ft (12-1/4 × 5.0 inch.)

Water added = 150 bbl required to fill annulus

The equivalent mud weight at TD will be:

a) 12.0 ppg

b) 10.5 ppg

c) 13.5 ppg

d) 11.0 ppg

210. What will be the free point constant and depth of pipe stuck using the given data?

Given data: 2-3/8 in. tubing-6.5lb/ft-ID = 2.441 in. ,25 in. of stretch with 20,0001b of pull force

a) 4530 and 5663 ft

b) 4530 and 5663 ft

c) 4230 and 5663 ft

d) 4430 and 5563 ft

211. The height (ft) of spotting fluid that will balance formation pressure in the annulus will be:

Given Data: Mud weight = 11.2 ppg

Weight of spotting fluid = 7.0 ppg

Amount of overbalance = 225.0 psi

a) 1030 ft

b) 1230 ft

c) 930 ft

d) 1530ft

212. Compute the pressure required to overcome the mud's gel strength inside the drill string with the data given below:

y = 101b/100 sq ft

d = 4.276in.

L = 12,000ft

a) 63.5psi

b) 93.5psi

c) 73.5psi

d) 83.5psi

213. The number of sacks of barite required to increase the density of 100bbl of 12.0 ppg (W_1) mud to 14.0 ppg (W_2) will be:

a) 40 sk/100 bbl

b) 240 sk/100 bbl

c) 340 sk/100 bbl

d) 140 sk/100 bbl

214. If the wellbore of 4000 m true vertical depth is filled up with 1.2 gm/cc mud and the frictional pressure loss in the annulus is 20 kg/cm^2, the equivalent circulating density will be:

a) 1.25 gm/cc

b) 1.5 gm/cc

c) 1.0 gm/cc

d) 2.0 gm/cc

215. The starting volume, bbl, of 12.0ppg (W_1) mud required to achieve 100 bbl (V_f) of 14.0 ppg (W_2) mud with barite will be

a) 85.5 bbl

b) 91.3 bbl

c) 80.5 bbl

d) 97.3 bbl

216. The number of barrels of water weighing 8.33 ppg (D_w) required to reduce 100 bbl (**V_1**) of 14.0 ppg (W_1) to 12.0 ppg (W_2) will be:

a) 54.5 bbl

b) 64.5 bbl

c) 74.5 bbl

d) 44.5 bbl

217. The volume of 11.0 ppg mud and 14.0 ppg mud required to build 300 bbl of 11.5 ppg mud will be:

Given: 300 bbl of 11.0 ppg mud on hand, and 300 bbl of 14.0 ppg mud on hand

a) 250 bbl of 11.0 ppg mud and 50 bbl of 14.0 ppg mud

b) 250 bbl of 14.0 ppg mud and 50 bbl of 11.0 ppg mud

c) 50 bbl of 11.0 ppg mud and 250 bbl of 14.0 ppg mud

d) 150 bbl of 11.0 ppg mud and 50 bbl of 14.0 ppg mud

218. Given: Proposed casing setting depth = 4000ft

Estimated fracture gradient = 14.2 ppg

Gas gradient = 0.12 psi/ft

Safety factor = 1.0ppg

Assume 100% of mud is blown out of the hole.

The maximum anticipated surface pressure (MASP), psi will be:

a) 1682psi

b) 2682psi

c) 4682psi

d) 3682psi

219. Leak-off pressure at casing setting depth (TVD) of 4000 ft was 1040 psi with 10.0 ppg in use. The maximum allowable shut-in casing pressure with a mud weight of 12.5 ppg will be:

a) 320 psi

b) 650 psi

c) 450 psi

d) 520 psi

Statement for Linked Answer Questions 220-221

The following data are given for a well while drilling:

Mud weight = 14.0

$\theta 300 = 37.0$

$\theta 600 = 64.0$

Hole diameter = 8.5inch

Pipe OD = 7.0 inch

220. The values of n and k will be:

a) 0.79 and 0.2684

b) 0.59 and 0.3684

c) 0.49 and 0.2684

d) 0.59 and 0.3684

221. The values of x, AVc, and GPMc for the above data will be:

a) 2035 t/min, 310, and 294 gpm

b) 2035 t/min, 210, and 294 gpm

c) 3035 t/min, 310, and 294 gpm

d) 1035 t/min, 310, and 294 gpm

222. A directionally drilled well requires that a correction be made in total drill collar weight because only a portion of the total weight will be available to the bit. Hence, compute this available partial weight when the total weight of collars is 45,000 lb and inclination angle is (I) is 25 degrees.

a) 49,7841b

b) 40,784 1b

c) 30,784 1b

d) 54,784 lb

223. A well is drilled at 8000 ft with 11.0 ppg mud and surface casing is set at 2500 ft. The hydrostatic pressure at the surface casing is---- psi

a) 1430 psi

b) 1000 psi

c) 2000 psi

d) 1600 psi

224. If a kick has been observed at 3000 m depth and the shut-in drill pipe pressure is 40 kg/cm^3 with 1.2 gm/cc in the wellbore, what mud weight is required to balance the formation pressure?

a) 1.33

b) 1.22

c) 1.44

d) None of the above

225. What is the cost per metre of drilling in a section of hole where the bit cost is Rs. 50,000 and the rig operating cost is Rs. 10, 000 per hour, the drilling time is 50 hours, the trip is 6 hours and the meterage drilled was 100m metre?

a) Rs. 61000 per m

b) Rs. 71000 per m

c) Rs. 81000 per m

d) Rs. 51000 per m

226. The minimum flowrate for a 12-1/4in. PDC bit will be:

a) 605.58 gpm

b) 505.87 gpm

c) 450.25 gpm

d) 480.35 gpm

Answers

Q1.	d	Q36.	c	Q71.	d	Q106.	d	Q141.	a
Q2.	d	Q37.	a	Q72.	d	Q107.	a	Q142.	c
Q3.	a	Q38.	a	Q73.	b	Q108.	a	Q143.	c
Q4.	d	Q39.	a	Q74.	d	Q109.	a	Q144.	b
Q5.	d	Q40.	c	Q75.	a	Q110.	c	Q145.	c
Q6.	d	Q41.	a	Q76.	c	Q111.	a	Q146.	d
Q7.	d	Q42.	a	Q77.	a	Q112.	c	Q147.	a
Q8.	c	Q43.	a	Q78.	c	Q113.	a	Q148.	b
Q9.	d	Q44.	a	Q79.	c	Q114.	c	Q149.	d
Q10.	d	Q45.	d	Q80.	a	Q115.	b	Q150.	a
Q11.	c	Q46.	a	Q81.	d	Q116.	c	Q151.	c
Q12.	d	Q47.	a	Q82.	d	Q117.	b	Q152.	b
Q13.	b	Q48.	a	Q83.	b	Q118.	c	Q153.	b
Q14.	b	Q49.	a	Q84.	c	Q119.	d	Q154.	c
Q15.	b	Q50.	a	Q85.	c	Q120.	d	Q155.	a
Q16.	d	Q51.	d	Q86.	d	Q121.	c	Q156.	d
Q17.	c	Q52.	a	Q87.	a	Q122.	b	Q157.	d
Q18.	c	Q53.	b	Q88.	c	Q123.	c	Q158.	c
Q19.	d	Q54.	c	Q89.	c	Q124.	d	Q159.	d
Q20.	a	Q55.	b	Q90.	b	Q125.	c	Q160.	a
Q21.	b	Q56.	a	Q91.	c	Q126.	d	Q161.	a
Q22.	b	Q57.	a	Q92.	b	Q127.	b	Q162.	b
Q23.	a	Q58.	a	Q93.	d	Q128.	c	Q163.	a
Q24.	a	Q59.	a	Q94.	d	Q129.	d	Q164.	a
Q25.	a	Q60.	a	Q95.	d	Q130.	a	Q165.	a
Q26.	a	Q61.	b	Q96.	d	Q131.	b	Q166.	b
Q27.	a	Q62.	a	Q97.	b	Q132.	a	Q167.	d
Q28.	b	Q63.	a	Q98.	a	Q133.	a	Q168.	c
Q29.	a	Q64.	a	Q99.	c	Q134.	d	Q169.	a
Q30.	c	Q65.	d	Q100.	c	Q135.	c	Q170.	a
Q31.	d	Q66.	a	Q101.	c	Q136.	a	Q171.	c
Q32.	b	Q67.	a	Q102.	a	Q137.	d	Q172.	a
Q33.	c	Q68.	a	Q103.	d	Q138.	d	Q173.	b
Q34.	c	Q69.	d	Q104.	b	Q139.	b	Q174.	c
Q35.	d	Q70.	d	Q105.	d	Q140.	d	Q175.	a

Answers									
Q176.	d	Q187.	b	Q198.	c	Q209.	a	Q220.	a
Q177.	a	Q188.	d	Q199.	a	Q210.	b	Q221.	d
Q178.	b	Q189.	c	Q200.	a	Q211.	a	Q222.	b
Q179.	d	Q190.	c	Q201.	d	Q212.	b	Q223.	a
Q180.	b	Q191.	b	Q202.	a	Q213.	d	Q224.	a
Q181.	d	Q192.	d	Q203.	c	Q214.	a	Q225.	a
Q182.	b	Q193.	a	Q204.	b	Q215.	b	Q226.	b
Q183.	a	Q194.	c	Q205.	a	Q216.	a		
Q184.	a	Q195.	b	Q206.	c	Q217.	a		
Q185.	b	Q196.	d	Q207.	a	Q218.	b		
Q186.	a	Q197.	b	Q208.	c	Q219.	d		

CHAPTER **3**

Reservoir Engineering

1. Which of the following is the measure of storage capacity that is capable of holding fluids?
   a) Porosity
   b) Permeability
   c) Formation volume factor
   d) None of the above

2. The percentage of interconnected pore space with respect to the bulk volume is called
   a) Total porosity
   b) Effective porosity
   c) Relative porosity
   d) None of the above

3. The void to grain ratio of sandstone sample is ¼. What is the porosity of this sample?
   a) 20%
   b) 25%
   c) 30%
   d) 40%

4. Matrix permeability originates at the time of deposition is called:
   a) Primary Permeability
   b) Secondary Permeability
   c) Tertiary Permeability
   d) None of the above

5. Results from the alteration of rock matrix by compaction, cementation, and solution is called
   a) Primary Permeability
   b) Secondary Permeability
   c) Tertiary Permeability
   d) None of the above

6. Permeability depends on
   a) internal surface of the pore space
   b) Distribution of the pore throat sizes
   c) Tortuosity of the fluid flow paths
   d) All of the above

7. Porosity of a rock depends on:
   a) Size and shape of the grains
   b) Nature of packing
   c) Presence and absence of vugs and fractures
   d) All of the above

8. The porosity of the cubic packing is:
   a) 47%
   b) 26%
   c) 35%
   d) 20%

9. The porosity of rhombohedral packing is:
   a) 32%
   b) 26%
   c) 12%
   d) 52%

10. The bulk density of a brine saturated core sample is 1.9 gm/c. If the density of brine solution and core matrix are 1.04 gm/cc and 2.55 gm/cc, the porosity of the formation will be:
    a) 36.5%
    b) 28.5%
    c) 24.5%
    d) None of the above

11. If 200 cubic feet of rock contains many tiny pores and spaces which together have a volume of 20 cubic feet, the porosity of the rock is
    a) 20%
    b) 10%
    c) 9.09%
    d) 11.11%

12. Which of the following well log can be used to determine porosity?
    a) Electrical resistivity log
    b) Neutron log
    c) Density log
    d) All of the above

13. A change in permeability value is observed while determining permeability with gases at different pressure. This is known as
    a) Klinkerberg effect
    b) Gravity effect
    c) Viscosity effect
    d) None of the above

14. Klinkerberg states that permeability to a gas is function of mean free path of the gas molecules. The factors which influence the mean free path are:
    a) Temperature, pressure and nature of the gas
    b) Temperature and Pressure of the gas
    c) Both (a) and (b)
    d) None of the above

15. The relative permeability can be expressed in
    a) Millidarcy,
    b) Square in
    c) Square cm
    d) Dimensionless Number

16. Which of the following equipment is normally used for the measurement of permeability?
    a) Ruska Air Permeameter
    b) Ruska Liquid Permeameter
    c) Hassler core holder
    d) All of the above

17. ________ is the permeability measured during flow of fluid through a single fracture or through a fracture network independent of surrounding rock.
    a) Intrinsic fracture permeability
    b) Absolute permeability
    c) Conventional fracture permeability
    d) None of the above

18. Absolute permeability of a rock saturated with oil and water is 100mD. If the relative permeability to water is 0.40, the effective permeability to oil is
    a) 60 mD
    b) 40 mD
    c) 50mD
    d) None of the above

19. The relative amount of a particular phase present in the pore space is called:
    a) Saturation
    b) Porosity
    c) Permeability
    d) None of the above

20. The maximum water saturation at which the water phase will remain immobile is known as :
    a) Critical water saturation
    b) Connate water saturation
    c) Irreducible water saturation
    d) All of the above

21. Which of the following statement is true about critical oil saturation?
    a) For the oil phase to flow, the saturation of the oil must exceed a certain value, which is termed critical oil saturation.
    b) At this particular saturation, the oil remains in the pores and, for all practical purposes, will not flow
    c) Both (a) and (b)
    d) None of the above

22. Which of the following statement is true about residual oil saturation?
    a) During the displacing process of the crude oil system from the porous media by water or gas injection (or encroachment), there will be some remaining oil left
    b) It is quantitatively characterized by a saturation value that is larger than the critical oil saturation
    c) Both (a) and (b)
    d) None of the above

23. Decreasing the saturation of the wetting phase is called
    a) Drainage
    b) Imbibition
    c) Wettability
    d) None of the above

24. Increasing the saturation of the wetting phase is called
    a) Drainage
    b) Imbibition
    c) Wettability
    d) None of the above

25. Which of the following is the basis for the classification of reservoir-aquifer systems?
    a) Degree of pressure maintenance
    b) Flow regimes and outer boundary conditions
    c) Flow geometries
    d) All of the above

26. Reservoir begins to fill up with oil is termed as
    a) Drainage
    b) Imbibition
    c) Both (a) and (b)
    d) None of the above

27. Reservoir begins to fill with water is termed as
    a) Drainage
    b) Imbibition
    c) Both (a) and (b)
    d) None of the above

28. What will be the effect on permeability when specific surface decreases?
    a) Decrease
    b) Increase
    c) Remains same
    d) None

29. The relationship characteristic of the displacement of the wetting phase from 100%saturation to the irreducible saturation is called
    a) Primary Drainage Curve
    b) Imbibition Curve
    c) Secondary Drainage Curve
    d) None of the above

30. The relationship characteristic of the displacement of the wetting phase from residual saturation to irreducible saturation is called
    a) Primary Drainage Curve
    b) Imbibition Curve
    c) Secondary Drainage Curve
    d) None of the above

31. The relationship characteristic of the displacement of the wetting phase from the irreducible saturation to the residual saturation is called
    a) Primary Drainage Curve
    b) Imbibition Curve
    c) Secondary Drainage Curve
    d) None of the above

32. Which of the following is true about the relationship between grain size and Sw irreducible?
    a) Small grain size - high Swirr
    b) Large grain size - low Swirr
    c) Both (a) and (b)
    d) None of the above

33. Sw irreducible is reached when
    a) K_{ro} is 100% and K_{rw} is 0%
    b) K_{ro} is 80% and K_{rw} is 20%
    c) K_{ro} is 50% and K_{rw} is 50%
    d) None of the above

34. Which of the following relationship is true?
    a) Large pore throats, high perm, low Sw irreducible, short transition zone
    b) Small pore throats, low perm, high Sw irreducible, Long transition zone
    c) Both (a) and (b)
    d) None of the above

35. As per Honarpour, Molina and Pi'son et al method, which formula is used for determination of relative permeability?
    a) kro = {(0.9 - Sw)/(0.9 - Swirr)}2

b) krw = {(Sw - Swirr)/(1 - Swirr)}3

c) both (a) and (b)

d) None of the above

36. The ratio of the actual path length to the straight path length is called

a) Tortuosity

b) Length ratio

c) Both of these

d) None of the above

37. Which of the following is true?

a) Swc ≤ Sw ≤ 1—Sor

b) Swc ≤ 1—Sor≤ Sw

c) Swc ≤ 1—Sor≤ Sw

d) None of the above

38. Which of the following statement is true?

a) Relative permeability is a function of fluid saturation

b) Since the effective k may range from zero to k, the relative k may have any value between zero and one

c) The sum of the relative permeabilities (kro + krg + krw) is always less than or equal to unity.

d) All of the above

39. Which of the following statement is true?

a) When both wetting and nonwetting phases flow together in a reservoir rock, each phase flows in separate and distinct paths. Their distribution results in a characteristic relative permeability curve

b) The practical range of water saturations is: Swc ≤ Sw ≤ 1—Sor

c) When Sw < Swc, the water will not flow and kw = 0

d) All of the above

40. Which of the following statement is true?

a) When Sw = 1, kw = k and rock is entirely saturated with water

b) When So < Soc, there will be no oil flow and ko = 0.

c) When So = 1, ko = k and rock is entirely saturated with oil

d) All of the above

41. Which of the following statement is true ?

a) The value of Sw at krw = 0 is called the Swc

b) The value of So at kro = 0 is called the Soc

c) When relative permeability of a phase is zero, there is still a considerable saturation of this phase

d) All of the above

42. The viscosity of crude oil measured at a pressure above bubble point and system temperature condition is known as

a) Under-saturated oil viscosity

b) Saturated oil viscosity

c) Plastic viscosity

d) Apparent viscosity

43. The viscosity of crude oil measured at a pressure above bubble point and system temperature condition is known as

e) Under-saturated oil viscosity

f) Saturated oil viscosity

g) Kinematic viscosity

h) Effective viscosity

44. Which of the following statement is true?

a) The non-wetting phase occupies the Larger pore spaces where flow occurs with the least difficulty

b) The non-wetting phase flows at the relatively lower saturation (Soc = 17%) than wetting phase (Swc= 20%).

c) The wetting phase occupies the smaller pore spaces, where ___ are the greatest and prevents it from flowing

d) All of the above

45. Which of the following statement is true?

a) Changes in the wetting phase saturation have only a small effect on the magnitude of the non-wetting phase relative permeability curve

b) At the low saturations, the wetting phase fluid occupies the small pore spaces and do not contribute to flow.

c) Imbibition relative permeability is displacement where the wetting phase saturation is increasing, i.e. water flood.

d) All of the above

46. Which of the following statement is true?

a) Drainage relative permeability is where the non-wetting phase saturation is increasing, i.e. gas expulsion of oil during primary depletion

b) The migration of oil is opposed by the capillary forces, which are functionally related to Pore size, Interfacial tension between oil and water and Adhesion of oil to mineral surfaces

c) Relative permeability is a function of fluid saturation

d) All of the above

47. Which of the following is true about Lorenz Process?

a) Sort thickness, porosities by interval perms in descending order

b) Compute the thickness products of (hk) and (hφ)

c) Normalize each by respective maximums

d) Plot normalized cumulative (hk) versus normalized cumulative (hφ)

48. ____ is the saturation above which the fluid will flow

a) Critical saturation, Sc

b) Residual oil saturation, Sr

c) Fluid unsaturation

d) None of the above

49. Below critical saturation, the fluid remains in the pores and, for all practical purposes, will

a) Not flow

b) Flow

c) Partially flow

d) None of the above

50. ____ is the saturation of oil that remains left in the pore space during the displacing process

a) Residual oil saturation, Sr

b) Critical saturation, Sc

c) Remaining saturation

d) None of the above

51. The values of flash liberation and differential liberation will not be equal for

a) $P \geq P_b$

b) $P < P_b$

c) $P = P_b$

d) $P=0$

52. In PVT behavior, solution gas liberated from the oil sample remains in contact with the oil surface to establish the equilibrium. The process is known as-

a) Flash liberation

b) Integrated liberation

c) Differential liberation

d) None of the above

53. Which of the following statement is true?

a) The amount of connate water present in the porous space varies from 100% below the oil zone to theoretically zero at heights above the free water level, known as transition zone

b) High Swc values are indicative of Small pore sizes and wells completed within this zone will produce hydrocarbons and water

c) The term residual saturation is usually associated with the non-wetting phase when it is being displaced by a wetting phase

d) All of the above

54. The distribution of various phases like oil, gas and water in the pore space depends on:

a) Saturation of the phase

b) Wettability to the rock

c) Both of these

d) None of the above

55. ____is the property of a fluid to adhere to a solid surface.

a) Wettability

b) Absorption

c) Adsorption

d) None of the above

56. The contact angle for non-wetting fluid is:

a) More than 90 degree

b) Less than 90 degree

c) Equal to 90 degree

d) None of the above

57. Relative permeability is:

a) ratio of effective permeability of a particular phase to the absolute permeability of the rock.

b) ratio of absolute permeability of a particular phase to the effective permeability of the rock

c) ratio of intrinsic fracture permeability of particular phase to the effective permeability of the rock

d) None of the above

58. Relative permeability of a rock to fluid phase is a function of:

a) Saturation

b) Wettability characteristics of the phase to the rock

c) Degree of consolidation of the rock

d) All of the above

59. Capillary pressure is a measure of the tendency of the rock to:

a) Suck in the wetting fluid phase and to repel the non-wetting phase

b) Suck in the non-wetting fluid phase and to repel the wetting phase

c) Suck in wetting fluid phase and non-wetting phase

d) Repel wetting fluid phase and non-wetting phase

60. The difference in pressure between two immiscible fluids across a curved interface at is called

a) Capillary Pressure

b) Atmospheric Pressure

c) Both of these

d) None of the above

61. Capillary force is affected by

a) Surface and interfacial tensions of the rock and fluids,

b) Pore size and geometry,

c) Wetting characteristics of the system and Fluid saturation

d) All of the above

62. If the pressure in oil is greater, rock is

a) water wet

b) Neutrally wet

c) oil wet

d) None of the above

63. If the pressure in water is greater, the rock is____

a) oil wet

b) Neutrally wet

c) water-wet

d) None of the above

64. When two fluids wet walls to the same extent and pressure of each fluid is the same, rock is

a) Oil wet

b) Water wet

c) Neutrally wet

d) None of the above

65. Which of the following statement is true?

a) Treatments that change the wettability of the formation from waterwet to oil-wet can significantly impair productivity

b) It is necessary to maintain pressure of the non-wetting fluid greater than that in the wetting fluid

c) Both (a) and (b)

d) None of the above

66. Leverett realized that capillary pressure depends on
   a) Porosity,
   b) Interfacial tension,
   c) Mean pore radius
   d) All of the above

67. The dimensionless function of saturation is called
   a) J-Function
   b) Z function
   c) K function
   d) None of the above

68. __________ is normally referred for those oils produced from deep reservoirs which occur in the reservoir as liquid and exhibit a large degree of shrinkage when brought to the surface.
   a) Volatile oil
   b) Black oil
   c) Non-volatile oil
   d) None of the above

69. Volatile oils are characterized by:
   a) Gravities > 45°API
   b) Formation factor > 2
   c) High solution gas ratios
   d) All of the above

70. The gradual changes in the production rate of a well may be caused by:
   a) Decreasing efficiency or effectiveness of the lifting equipment
   b) Reduction of productivity index or completion factor or increase in skin effect
   c) Changes in bottom hole pressure, gas-oil ratio, and water percentage
   d) All of the above

71. Tortuosity (t) can be determined by determining formation resistivity factor (F) and porosity by the using the following relationship.
   a) $t = F*ø$
   b) $t^2 = F*ø$
   c) $t = F/ø$
   d) $t^2 = F/ø$

72. Which of the following statement is true about the wet gas reservoir?
   a) It contains colorless gas
   b) It primarily contains methane and ethane
   c) Its initial GOR is 20,000 – 100,000 scf/stb
   d) All of the above

73. Which of the following statement is true about the gas condensate reservoir?
   a) Its initial GOR is 6000 – 20000 scf/stb
   b) It primarily contains C_{7+}
   c) The API° gravity of fluid is 45-65
   d) All of the above

74. The primary reservoir characteristics are
   a) Type and number of fluids in the reservoir
   b) Flow regimes
   c) Reservoir geometry
   d) All of the above

75. You are calculating the original gas in place using the volumetric method. Which of the following has the least influence on the final result?

a) Connate water saturation
b) Initial reservoir pressure
c) Permeability
d) Porosity

76. Which of the following is the least likely source for water which encroaches into a reservoir as pressure declines?

a) Artesian flow
b) Compressibility of the rock in the aquifer
c) Expansion of the water in the aquifer
d) Water coming out of solution from the oil as the pressure drops

77. Which of the following criteria will be used for the determination of optimum number of wells?

a) There should be minimum interference between the wells
b) Well should be spaced in such a manner that they could be utilized in future for pressure maintenance or for EOR
c) The well should be placed considering isopay map at the place where effective thickness is goof and well may pay its cost.
d) All of the above

78. For decline-curve analysis to be accurate, several conditions must be satisfied.

Which of the following conditions is not appropriate?

a) Drainage area of the well remains constant with time.
b) Skin factor of well is not changing with time.
c) Well is in transient flow regime.
d) Well produces at constant BHP.

79. Volumetric estimate of original gas in place for gas reservoir A was 100 Bscf. A material-balance estimate, using a **p/z** plot, was 50 Bscf. Which of the following statements is most likely true?

a) One or more producing wells intended to produce from Reservoir A is probably completed by mistake in another non-communicating reservoir.
b) Some of the mapped gas may not be in pressure communication with the producing wells.
c) The estimate based on the **p/z** plot cannot possibly be correct.
d) The reservoir almost certainly has water influx.

80. In a low-permeability gas formation, the deliverability curve determined from a plot of **Δp2** vs. **q**, using available test data from a four-point backpressure test with equal-length flow periods, will usually not be "stabi-

lized." Therefore, the ability of the gas well to deliver gas at a given backpressure will be:

a) About the same as predicted from the deliverability curve plotted using available test data.

b) Greater than predicted from the deliverability curve plotted using available test data.

c) Less than predicted from the deliverability curve plotted using available test data.

d) Uncorrelated to the value predicted from the deliverability plotted using available test data.

81. An oil reservoir was originally undersaturated and was known to have no water influx. The field is to be waterflooded and part of the preflood strategy is to determine the new bubble point for this state of depletion. The initial reservoir pressure is 3,000 psia with an initial oil saturation; Soi = 0.90. The abandonment pressure is 1,200 psia. At the abandonment pressure the oil formation volume factor is 1.20. Cumulative oil production to the bubble point is 250,000 stb. Using the above production information and saturations, assume Bob = 1.460 rb/stb and the Boi = 1.441 rb/stb and that rock and water expansion are negligible. What is the new bubble point solution gas oil ratio, scf/stb that would be indicated by increasing the reservoir pressure with water injection such that all the free gas was driven back into the solution.

a) 465

b) 512

c) 816

d) 935

82. What is the original oil in place in stb of a reservoir with the following parameters?

Area = 80 acres, net pay thickness = 50 ft

Average porosity = 20%,

Initial water saturation, Swc = 10%,

Initial gas saturation, Sg = 10%,

Initial Oil Formation Volume Factor, Boi = 1.42 rb/stb,

Initial Gas Formation Volume Factor, Bgi = 0.8 rb/mscf.

a) 3,496,563

b) 3,933,633

c) 6,206,400

d) 300,000,000

83. Using the following parameters, what is the ratio of the gas cap to the oil rim, m?

Area = 80 acres,

net pay thickness = 50 ft

average porosity = 20%,

initial water saturation, Swc = 10%,

initial gas saturation, Sg = 10%,

Initial Oil Formation Volume Factor, Boi = 1.42 rb/stb,

Initial Gas Formation Volume Factor, Bgi = 0.8 rb/mscf.

a) 0.125

b) 0.22

c) 0.5

d) Not enough information

84. A producer-injector pair of wells in an infinite-acting reservoir has been producing two weeks. The distance between the wells is 500 feet Assume that the injection and production rates are equal at 1,500 bpd and that the rock and fluid properties are the same for both injector and producer. Reservoir Properties for this formation are:

Permeability = 250 md

Porosity = .22

Net thickness = 30 ft

Oil viscosity = .5 cp

Water viscosity = .5 cp

Total compressibility = 4E-5 psi

Injector Skin = 10

Producer Skin = 0

Wellbore radius of both well is 5.5 inches

Bo = Bw = 1.0 rb/stb

The expected pressure drop in psi, in the production well will be:

a) 26.91

b) 98.78

c) 239.98

d) 656.42

Statement for Linked Answer Questions 85 and 86

The following information are given for an oil well:

Reservoir pressure: 6000 psi

Bottom hole flowing pressure: 5500 psi

Oil viscosity: 0.25 cp

Permeability of the reservoir: 20 md

Pay zone thickness: 30 ft

Drainage radius: 1000 ft

Wellbore radius: 6 in.

85. What is the oil production rate?

a) 1117.8 bbl/day

b) 2117.8 bbl/day

c) 3117.8 bbl/day

d) None of the above

86. If by applying an artificial lift, the bottom hole pressure is reduced to 3000 psi, at what rate oil will produce?

a) 6707 bbl/day

b) 7707 bbl/dy

c) 8707 bbl/day

d) 9707 bbl/day

87. Which of the following statements is true regarding undersaturated oil reservoirs?

a) The two phase and single phase factors are equal above the bubble point

b) The initial solution gas-oil ratio does not equal the solution gas-oil ratio at a specified pressure above the bubble point

c) Below the bubble point, the two-phase factor decreases while pressure decreases

d) Below the bubble point, the single phase factor increases while pressure decreases

88. Which of the following is the least likely characteristic of a reservoir fluid?

a) Compressible liquid
b) Gas
c) Incompressible liquid
d) Superfluid

89. The following informations are given for a oil reservoir: 160 acres, 20 ft net pay, 10% porosity, 25% water saturation, FVF =1.0 and 35% recovery factor The amount of oil reserve would be ---
a) 651672 bbls
b) 605172 bbls
c) 321762 bbls
d) None of the above

90. Reservoir simulators are well-suited to help us to determine
a) Aquifer size and strength, remaining reserves, corporate overhead
b) Future water production, optimal infill well location, results of fracture stimulation treatments
c) Optimal well spacing in reservoirs, coning characteristics of wells, and optimal flood patterns in improved recovery projects
d) Unique reservoir description, optimal enhanced recovery method, good matches of reservoir history

91. The factors that cause recovery factors to vary in uncertain ways in oil reservoirs do not include
a) Distribution of permeabilities vertically and horizontally
b) Presence and location of reservoir boundaries
c) Presence and location of reservoir heterogeneities
d) Unexpected variations in oil prices

92. A reservoir produces gas at an initial gas/liquid ratio (GLR) of 20,000 Scf/STB. After some period of production at constant GLR, the producing GLR begins to increase with time. Essentially all the produced liquids are gas condensate.
The reservoir fluid is most likely which one of the following types?
a) Black oil
b) Dry gas
c) Gas condensate (retrograde gas)
d) Wet gas

93. The proper ranking of average (typical, not exceptional) oil reservoir recovery efficiency
(from lowest to highest) by drive mechanism is
a) solution-gas drive; rock-and-fluid expansion drive; water drive; expanding gas-cap drive
b) solution-gas drive; expanding gas-cap drive; water drive; gravity drainage drive
c) rock-and-fluid expansion drive; solution-gas drive; water drive; expanding gas-cap drive
d) rock-and-fluid expansion drive; expanding gas-cap drive; gravity-drainage drive; water drive

94. Low-permeability or tight gas reservoirs ordinarily produce at economic rates because they are

a) Stimulated with long, highly conductive hydraulic fractures

b) Provided substantial tax and other economic incentives by governmental authorities

c) Drilled with very wide well spacing, often 640 acres (2.59 × 106 m2), and drain large areas efficiently with limited numbers of vertical wells

d) Completed in formations of massive thickness

95. An oil reservoir was originally undersaturated and was known to have no water influx. The field is to be waterflooded and part of the preflood strategy is to determine the new bubble point for this state of depletion. The initial reservoir pressure is 3,000 psia (with an initial oil saturation; Soi = 0.90. The abandonment pressure is 1,200 psia . At the abandonment pressure the oil formation volume factor is 1.20. Cumulative oil production to the bubble point is 250,000 stb. Using the above production information and saturations, assume Bob = 1.460 rb/stb and the Boi = 1.441 rb/stb and that rock and water expansion are negligible. If the new solution gas oil ratio is 750 scf/stb, what is the new bubble point pressure, psia (KPa), nearest?

a) 3,000

b) 2,160

c) 1,800

d) 1,200

96. A high-pressure gas reservoir had a discovery pressure of 10,000 psia. PVT analysis of the gas indicated the following variation in deviation factor, **z**, with pressure.

p, psia

z, dimensionless 10,000 (6.895 × 104 KPa) 1.47

8,000 (5.516 × 104 KPa) 1.30

Formation ater compressibility was 3.5 × 10-6 psia-1, initial water saturation 25%, and constant formation compressibility 2 × 10-5 psia^{-1}. Estimate the percentage, %, of original gas in place produced when reservoir pressure drops from discovery pressure to 8,000 psia.

a) 9.54

b) 14.6

c) 20.0

d) 85.4

97. The assumption not used in the derivation of the radial flow form of the diffusivity equation is:

a) Uniform thickness across the reservoir

b) A fluid with small and constant compressibility

c) Homogeneous and isotropic medium

d) Flow into the wellbore continues after the well is shut in

98. Which influences the compressibility factor of a gas?

a) Temperature

b) Pressure

c) Composition

d) All of the above

99. A reservoir fluid has formation volume factor 1.2417 at Reservoir Pressure (P_r) above Bubble Point pressure. With decline pressure of reservoir towards Bubble Point, the Formation Volume Factor of Oil will-

a) Decrease till 1

b) Increase

c) Decrease till 0

d) Remains same

100. Reserves for high-pressure volumetric dry gas reservoirs are often obtained by extrapolating a plot of p/z vs. Gp to the abandonment point. Equivalent results could be obtained by plotting:

a) 1/Bg vs. Gp

b) p vs. Gp

c) p2 vs. Gp

d) Bg vs. Gp

101. A dry gas reservoir has an estimated bulk volume of 9,418 acre-feet, an average porosity of 0.20, and an initial water saturation of 0.35. If the gas formation volume factor is 0.004503 ft3/scf, the initial gas in place (MMscf) is most nearly:

a) 211

b) 6,380

c) 11,800

d) 18,200

102. A new oil well is expected to produce at the allowable rate of 250 STB/D for the first 18 months and then start a constant percentage decline of 5% per month. Using this scenario, the total production of oil (thousand STB) during the second year of the productive life would be most nearly:

a) 40

b) 68

c) 85

d) 176

103. An oil lease is currently producing at a rate of 450 STB/D. The rate has been declining at a constant percentage of 30% per year (loss ratio = 0.3). If the economic limit on production is 60 STB/D, the remaining productive life (months) for the lease would be most nearly:

a) 5.6

b) 29

c) 68

d) 81

104. A dry gas reservoir had initially 11,844 MMscf of gas in place with a formation volume factor of 0.004503 res ft3/scf. 6,097 MMscf of gas has been produced along with 1,205 STB of water (Bw = 1.035 res ft3/scf). The present formation volume factor is 0.004896 res ft3/scf. The volume of water influx (million ft3) at this time is most nearly:

a) 6.11

b) 18.2

c) 25.2

d) 32.2

105. A waterflood is planned for a depleted oil reservoir having a bulk volume of 1,289 acre-feet, a porosity of 0.22, and an initial (connate) water saturation of 0.26. The reservoir has produced oil and gas (no water) to a point where the average gas saturation is 0.11. Formation volume factors for oil and water are 1.36 and 1.0 respectively, and the residual oil saturation to water is 0.28. Assuming a residual gas saturation of 0 and no oil production before fillup, the volume of water (thousand STB) necessary to reach liquid fill up in the reservoir is most nearly:

a) 180

b) 240

c) 560

d) 770

106. A commercial oil reservoir has recently been discovered. If water influx is present but ignored in early material balance calculations of original oil in place, the calculated value for OOIP would most likely be:

a) too high

b) too low

c) correct

d) independent of water influx

107. A commercial oil reservoir has recently been discovered. If the oil reservoir is producing at pressures above the bubble point pressure, the factor in material balance calculations that is far more important than if the reservoir were producing below the bubble point pressure is:

a) Water influx

b) Dip angle of the reservoir

c) Produced gas/oil ratio

d) Formation compressibility

108. For a well in a dry gas reservoir you have the following information:

Well spacing 500 acres

Net sand thickness 50 feet

Porosity 0.15

Initial water saturation 0.22

Reservoir temperature 189°F

Reservoir pressure 5,257 psia

Gas deviation factor (at 5,257 psia) 1.015

The original gas in place (MMscf) is most nearly:

a) 7,000

b) 33,000

c) 36,000

d) 46,000

109. A volumetric (no water influx) dry gas reservoir had an initial pressure of 4,750 psia with a gas deviation factor of 0.975. At an abandonment reservoir pressure of 1,500 psia with a gas deviation factor of 0.892, the gas recovery factor (%) would be most nearly:

a) 34.5

b) 65.5

c) 68.4

d) 71.1

110. A gas well is producing with a sand face pressure above the dew point pressure. The surface producing gas/liquid ratio is 15,000 scf of surface gas (γg = 0.67) per STB of condensate (γo

= 0.759, Mo = 124). The specific gravity of the produced reservoir gas (fraction, air = 1.0) is most nearly:

a) 0.675
b) 0.705
c) 0.855
d) 1.015

111. A dry gas reservoir had an original gas in place of 12,000 MMscf at a pressure of 3,500 psia

(Bgi = 0.86625 M res bbl/ MMscf). Cumulative production to a reservoir pressure of 2,500 psia is 3,500 MMscf of gas (Bg = 1.17275 M res bbl/MMscf) and 100 MSTB of water (Bw = 1.04 res bbl/STB). The cumulative water influx (M res bbl) during this period is most nearly:

a) 410
b) 530
c) 2,510
d) 2,980

112. A currently depleted and shut-in oil reservoir originally contained 10,000 (thousand STB) of undersaturated oil with a formation volume factor of 1.3389 res bbl/STB and a connate water saturation of 0.35. Cumulative oil production to date has been 937.5 (thousand STB) of oil. The oil formation volume factor is now 1.2761 res bbl/ STB. Assuming no water influx, the current oil saturation is most nearly:

a) 0.30
b) 0.56
c) 0.60
d) 0.62

113. You have recently drilled and logged an exploratory well in a new basin. Based on the open-hole logs and a drillstem test, you suspect the prospective formation from 7,030 ft to 7,300 ft contains an oil/water contact. A repeat formation tester has been run and the following pressures were recorded:

Depth, ft	Pressure, psi
7,240	3,258
7,190	3,235
7,140	3,216
7,090	3,199
7,040	3,182

The sample chamber contained an insufficient amount of oil to determine its properties. The specific gravity of the water was measured to be 1.07. The depth (ft) of the oil/water contact is most nearly:

a) 7,140
b) 7,165
c) 7,175
d) 7,190

114. You have just drilled a well on the down-thrown side of a large fault to test whether the main producing interval in the field is productive on this side of the fault. The following parameters are known from the existing production in the field:

Water resistivity 0.018 ohm-meters

Invading fluid density 1.0 g/cm3

a = 0.625; m = 2.15; n = 2

If the porosity in the zone of interest is 9.4% and the

true formation resistivity is 30 ohm-meters, the percent hydrocarbon saturation is most nearly:

a) 94
b) 87
c) 75
d) 25

115. You have just drilled a well on the down-thrown side of a large fault to test whether the main producing interval in the field is productive on this side of the fault. The following parameters are known from the existing production in the field:

Water resistivity 0.018 ohm-meters

Invading fluid density 1.0 g/cm3

Oil formation volume factor 1.15 RB/STB

a = 0.625; m = 2.15; n = 2

If the porosity in the zone of interest is 9.4% and the true formation resistivity is 20 ohm-meters, the original oil in place (STB/ac-ft) is most nearly:

a) 444
b) 510
c) 1,228
d) 2,866

116. The grain volume of the cylindrical sample was measured in a two cell Boyle's law porosimeter. The porosity of the sample for the following given data is:

Sample dimensions:

Length = 4.00 cm

Diameter = 2.50 cm

Porosimeter Data:

V_1 = 25.0 cc

V_2 = 50.0 cc

P_1 = 100.0 psig

P_2 = 50.0 psig

a) 21.6%
b) 27.5%
c) 21.5%
d) 15.0%

117. The bulk volume of a core sample was measured by mercury displacement as 25.0 cc.

Given:

W_d = 50.25 gm

W_s = 54.50 gm

= 0.701 gm/cc

The porosity and grain density of the saturated sample are

a) 28.5% and 2.65 gm/cc
b) 24.2% and 2.65 gm/cc
c) 32.5% and 2.2 gm/cc
d) 24.2% and 2.65 gm/cc

Statement for Linked Answer Questions 118-121

The following information are given on a core sample:

1) Sample weight as received from field = 53.50 gm
2) Water volume recovered during extraction = 1.50 cc
3) Sample weight after extracting and drying = 51.05 gm
4) Density of core oil = 0.850 gm/cc
5) Bulk volume of sample = 23.60 cc
6) Grain density of sample = 2.63 gm/cc compute the porosity, and the oil, water, and gas saturations.

118. The porosity of the sample is
   a) 32.0%
   b) 15.0%
   c) 10.5%
   d) 17.8%

119. The oil saturation of the sample is
   a) 27%
   b) 35%
   c) 42%
   d) 12%

120. The water saturation of the sample is
   a) 36%
   b) 25%
   c) 42%
   d) 12%

121. The gas saturation of the sample is
   a) 47%
   b) 37%
   c) 27%
   d) 57%

Statement for Linked Answer Questions 122-124

A well is producing from a reservoir having the relative characteristics of two phase flow behavior. The following data are available:

$p_e = 2500$ psi
$\mu_o = 5.0$ cp
$p_w = 1000$ psi
$\mu_w = 0.6$ cp
$r_e = 700$ ft
h = 25 ft
$r_w = 0.33$ ft
k =50 md (absolute)
$B_o = 1.30$ (formation volume factor)
$k_{ro} = 0.23$ at $S_o = 0.40$
$K_{ro} = 0.45$ at $S_o = 50\%$

122. What will the steady state tank oil production rate be if the water saturation is at the critical value?
   a) 2.70 bbl/day
   b) 5.70 bbl/day
   c) 8.70 bbl/day
   d) 10.70 bbl/day

123. What will the tank oil flow rate be when $S_o = 0.50$, $p_e - p_w = 1000$ psi, $\mu_o = 7.0$ cp, and $B_o = 1.20$?
   a) 62 bbl/day
   b) 72 bbl/day
   c) 78 bbl/day
   d) 92 bl/day

124. What will the producing water-oil ratio be when $S_o = 0.40$ and $\mu_o = 7.5$ cp?
   a) 25
   b) 35
   c) 29
   d) 15

Statement for Linked Answer Questions 125-127

Given that a well has an initial production rate of 1130 BOPD and declined to 1050 BOPD after 3 months of continuous production:

125. Predict the production rate after a further 9 months using nominal exponential decline.
   a) 842.4 BOPD
   b) 742.4 BOPD
   c) 642.4BOPD
   d) 942.4 BOPD

126. Predict the production rate after a further 9 months using Nominal hyperbolic decline (b = 0.5)

a) 955.1 BOPD

b) 855.1 BOPD

c) 755.1 BOPD

d) 655.1 BOPD

127. Predict the production rate after a further 9 months using Nominal harmonic decline

a) 866.1 BOPD

b) 766.1 BOPD

c) 966.1 BOPD

d) 666.1 BOPD

128. Which reservoir drive mechanism has the highest recovery factor?

a) water drive

b) Gas cap

c) Solution drive

d) Combination drive

129. What type of reservoir drive maintains a constant reservoir pressure most of the production life?

a) Water drive

b) Gas cape

c) Solution drive

d) Combination drive

130. What is the energy called that produces the oil and gas?

a) Reservoir Drive

b) Solution drive

c) Aquifer

d) None of the above

131. Name the order of the three (3) drives (water, dissolved and free gas) from highest to lowest decline curve slope?

a) Dissolved, free gas, water drive

b) Free gas, water, dissolved drive

c) Dissolved, water, free gas drive

d) None of the above

132. What is the name for reserves that are calculated based on tests that the oil can be produced with reasonable certainity?

a) Proven

b) Probable

c) Possible

d) None of the above

133. What type of reservoir drive has a low initial gas/oil ratio increasing to a max near the end of production?

a) Dissolved gas

b) Gas cap drive

c) Water drive

d) None

134. Name the type of reservoir drive that has a constant oil/gas ratio most of the producing time increasing near the end of production?

a) Free gas cap

b) Water drive

c) Gas cap drive

d) Solution gas drive

135. A decline curve compares these two parameters?
   a) Production and time
   b) Production and permeability
   c) Production and saturation
   d) None of the above

136. What will be the gas reserves based on the given data:(320 acres, 12 ft net pay, 15% porosity, 15% water saturation, Bg= 0.00875 reservoir cf/scf and 70% recovery factor)?
   a) 1.71 bscf
   b) 1.85 bscf
   c) 2.95 bscf
   d) 3.22 bscf

137. Which of the following curve determines the economic limit for a well?
   a) Decline curve
   b) Saturation curve
   c) Permeability curve
   d) None of the above

138. What will be the oil reserves based on the given data: (160 acres, 20 ft net pay, 10% porosity, 25% water saturation, FVF=1.0 and 35% recovery factor)?
   a) 651,672 bbls
   b) 751,225 bbls
   c) 552,365 bbls
   d) 600,754 bbls

139. What type of reservoir drive maintains a constant reservoir pressure most of the production life?
   a) Water drive
   b) Depletion drive
   c) Gas cap drive
   d) None of the above

140. The recovery factor is:
   a) A ratio of the OOIP to the amount of oil in the reservoir at the economic limit
   b) A ratio of the OOIP to the amount of gas in the reservoir at the economic limit
   c) A ratio of the OOIP to the amount of total reserves at the economic limit.
   d) None of the above

141. What is the boundary water called that moves in a water drive reservoir?
   a) Connate water
   b) Water encroachment
   c) Bound water
   d) None of the above

142. Which two factors influence the gas formation volume factor?
   a) Permeability and Saturation
   b) Permeability and Pressure
   c) Pressure and Temperature
   d) None of the above

143. The decline curve for which type of reservoir has very steep initial production?
   a) Fractured reservoir
   b) Un-fractured reservoir
   c) Both (a) & (b)
   d) None of the above

144. The shrinkage factor for an oil is based on:
   a) the oil and gas ratio
   b) Specific gravity of oil
   c) Viscosity of oil
   d) None of the above

145. The unit of measure of permeability is:
   a) Darcy
   b) Centistokes
   c) Centipoise
   d) N/m^2

146. Ability to allow fluid to pass through a material is called
   a) Permeability
   b) Porosity
   c) Viscosity
   d) None of the above

147. Which of the following assumption is used in deriving the equation for fluid flow in porous media?
   a) Flow is single phase and laminar
   b) medium is homogenous
   c) All of the above
   d) None of the above

148. Non-Associated Gas ________
   a) do not contain significant quantities of crude oil
   b) contains significant quantities of crude oil
   c) contains sulphur dioxide gas
   d) None of the above

149. Associate gas refers to the situation:
   a) when gas dissolved into crude oil and no free gas is present
   b) When gas do not contain significant amount of crude oil
   c) When gas dissolved in the crude oil and some amount of free gas is present
   d) None of the above

150. Which of the following statement is true?
   a) The recovery factor of gas cap drive is smaller than that of water drive
   b) The recovery factor of water drive is smaller than that of gas cap drive
   c) The recovery factor of gas cap drive is same as that of water drive
   d) None of the above

151. The law which quantifies the rate of flow for fluids through porous media via the concept of permeability is
   a) Darcy's Law
   b) Newton's law
   c) Navier stoke's law
   d) None of the above

152. ________is commonly used to describe diffusion in porous media
   a) Tortuosity
   b) Porosity
   c) Saturation
   d) Wettability

153. The reservoir rocks which contain hydrocarbons that can be recovered is known as

a) pay zone
b) Well
c) Both (a) and (b)
d) None of the above

154. Type of drive when the hydrocarbons are initially above the bubble point is known as

a) solution gas drive
b) gas cap drive
c) water drive
d) None of the above

155. Type of drive when the hydrocarbons are initially free gas accumulated at the top of the reservoir and receives energy from the high compressibility of the gas is known as

a) Gas cap drive
b) Water drive
c) Solution gas drive
d) None of the above

156. Type of drive when the hydrocarbons are segregated by the natural density and can be responsible for moving the fluids to the wellbore is known as

a) gravity drive
b) Gas cap drive
c) Water drive
d) Solution gas drive

157. Type of drive when the hydrocarbons are overburden with stress making the reservoir pressure is reduced is known as

a) Compaction drive
b) Gas cap drive
c) Water drive
d) Solution gas drive

158. Type of drive when the hydrocarbons are in contact with a large volume of water which aids in the production by driving the energy is known as

a) water drive
b) Gas cap drive
c) Water drive
d) Solution gas drive

159. Type of drive when the hydrocarbons are enclosed by a porous media and the only energy comes from the pressure is known as

a) depletion drive
b) Gas cap drive
c) Water drive
d) Solution gas drive

160. The state of pressure and temperature at which all intensive properties are equal is called

a) Critical Point
b) Bubble-point
c) Dew-point
d) None of the above

161. The maximum temperature above which liquid cannot be formed regardless of pressure is called

a) Criconden therm
b) Critical temperature
c) Bubble point temperature
d) None of the above

162. The maximum pressure above which no gas can be formed regardless of regardless of temperature is called
    a) Bubble point pressure
    b) Critical pressure
    c) Criconden bar
    d) None of the above

163. The line separating the liquid-phase region from the two-phase region is known as
    a) Bubble-point curve
    b) Dew-point curve
    c) Liquid phase curve
    d) None of the above

164. The line separating the vapor-phase region from the two-phase region is known as
    a) Bubble-point curve
    b) Dew-point curve
    c) Vapour phase curve
    d) None of the above

165. The pressure at which a bubble of gas is first liberated from the oil is called
    a) Bubble-Point Pressure
    b) Liberation pressure
    c) Dew point pressure
    d) None of the above

166. ________is the temperature at which a gas becomes a liquid is called
    a) Bubble point
    b) Dew point
    c) Boiling point
    d) None of the above

167. What are the formations required for a volatile, wet gas and dry gas reservoir?
    a) The upper section of the phase diagram and the critical point
    b) Cricondenbar
    c) Cricondentherm
    d) All of the above

168. In under-saturated oil reservoir, initial reservoir pressure is
    a) greater than bubble point pressure of the fluid
    b) less than bubble point pressure of the fluid
    c) equal to bubble point pressure of the fluid
    d) None of the above

169. In saturated oil reservoir, initial reservoir pressure is
    a) greater than bubble point pressure of the fluid
    b) less than bubble point pressure of the fluid
    c) equal to bubble point pressure of the fluid
    d) None of the above

170. In gas cap reservoir, initial reservoir pressure is
    a) greater than bubble point pressure of the fluid
    b) less than bubble point pressure of the fluid
    c) equal to bubble point pressure of the fluid
    d) None of the above

171. An accumulation of natural gas at the top of a crude oil reservoir which provides the pressure to rapidly evacuate the crude oil from the reservoir is known as

a) Gas cap
b) Bubble cap
c) (b) & (c)
d) None of the above

172. A petroleum reservoir has a formation pressure of 175 bar. The bubble point pressure of oil is 100 bar, the reservoir will be-

a) At critical point
b) Under saturated
c) Saturated
d) Over saturated

173. The ratio of the volume of oil (plus the gas in solution) at the prevailing reservoir temperature and pressure to the volume of oil at standard conditions BBL/STB is called

a) oil formation volume factor
b) Total formation volume factor
c) Gas formation volume factor
d) None of the above

174. The ratio of the total volume of the hydrocarbon mixture (i.e., oil and gas, if present), at the prevailing pressure and temperature per unit volume of the stock-tank oil bbl/stb is called

a) oil formation volume factor
b) Total formation volume Factor
c) Gas formation volume factor
d) None of the above

Statement for Linked Answer Questions 173-176

An oil reservoir exists at its bubble-point pressure of 3000 psia and temperature of 160°F. The oil has an API gravity of 42° and gas-oil ratio of 600 scf/STB. The specific gravity of the solution gas is 0.65. The following additional data are also available:

Reservoir area =640 acres,
Average thickness =10 ft,
Connate water saturation =0.25,
Effective porosity =15%

175. The specific gravity of the oil will be:

a) 0.8859
b) 0.9231
c) 0.8935
d) 0.8156

176. Initial oil formation volume factor using standing equation will be:

a) 1596 bbl/STB
b) 1396 bbl/STB
c) 1896bbl/STB
d) 1682 bbl/STB

177. Pore volume will be :

a) 7447680 bbl
b) 8462720 bbl
c) 6543210bbl
d) 5467890bbl

178. Initial oil in place will be:

a) 4921670STB
b) 4276998 STB
c) 5275690STB
d) 6231450STB

179. The gas oil ratio for ordinary black oil is

a) between 200-700 scf/stb

b) less than 200 scf/stb

c) 2000-3200 scf/stb

d) than 3000 scf/stb

180. The gas oil ratio for low-shrinkage (volatile) crude oil is

a) between 200-700 scf/stb

b) less than 200 scf/stb

c) 2000-3200 scf/stb

d) than 3000 scf/stb

181. The gas oil ratio for high-shrinkage (volatile) crude oil:is

a) between 200-700 scf/stb

b) less than 200 scf/stb

c) 2000-3200 scf/stb

d) than 3000 scf/stb

182. The gas oil ratio for near-critical crude oil is

a) between 200-700 scf/stb

b) less than 200 scf/stb

c) 2000-3200 scf/stb

d) than 3000 scf/stb

183. The oil formation volume factor for low shrinkage crude oil is

a) less than 1.2 bbl/stb

b) less than 2 bbl/stb

c) greater than 2 bbl/stb

d) None of the above

184. The oil formation volume factor for near critical crude oil is

a) less than 1.2 bbl/stb

b) less than 2 bbl/stb

c) greater than 2 bbl/stb

d) None of the above

185. The oil formation volume factor for high shrinkage crude oil is

a) less than 1.2 bbl/stb

b) less than 2 bbl/stb

c) greater than 2 bbl/stb

d) None of the above

186. API gravity of ordinary black oil is

a) 15-40 °API

b) less than 35 °API

c) 45-55 °API

d) None of the above

187. API gravity of low-shrinkage (volatile) crude oil is

a) 15-40 °API

b) less than 35 °API

c) 45-55 °API

d) None of the above

188. API gravity of high-shrinkage (volatile) crude oil is

a) 15-40 °API

b) less than 35 °API

c) 45-55 °API

d) None of the above

189. Which of the following statement is true?

a) Gas formation volume factor is used to relate the volume of gas, as measured at reservoir conditions, to the volume of the gas as measured at standard conditions

b) Volume of 1 lb of gas at surface is 379.4 scf/lb-mol
c) Both (a) and (b)
d) None of the above

190. Which of the following parameters influence oil viscosity?
a) Temperature
b) Pressure
c) Oil gravity
d) All of the above

191. Which of the following is incompressible fluid?
a) Dead oil
b) Live oil
c) Gas
d) None of the above

192. Which of the following is slightly compressible fluid?
a) Dead oil
b) Live oil
c) Gas
d) None of the above

193. Which of the following is compressible fluid?
a) Dead oil
b) Live oil
c) Gas
d) None of the above

194. Which of the following is true about sampling?
a) Samples must be representative
b) Sampling must be reproducible
c) Sampling must be repeatable.
d) All of the above

195. For volatile oils, which sampling method is preferred?
a) Subsurface sampling
b) Wellhead sampling
c) Surface separator sampling
d) None of the above

196. Which of the following is true about reservoir simulation?
a) The multidimensional model to perform material balance in discrete volumes of the reservoir is called reservoir model
b) It estimates of OOIP and RF
c) It evaluate of alternate reservoir management strategies
d) All of the above

197. Which of the following is a type of reservoir simulation?
a) Black oil
b) Thermal
c) Compositional
d) All of the above

198. Which of the following is true about black oil simulation?
a) Assumes constant composition for each phase (oil, gas and water) and assumes isothermal conditions
b) Fluid and rock properties function of pressure only
c) Considers gravitational and capillary forces
d) All of the above

199. Which of the following is true about Compositional simulation?
   a) Consider several components, which could be in different phases
   b) Composition is determined based on equations of state or equilibrium constants
   c) Fluid properties depend on composition and pressure
   d) All of the above

200. Which of the following is true about Thermal simulation?
   a) Includes an energy balance equation
   b) Fluid properties depend on pressure, temperature and/or composition
   c) Both (a) and (b)
   d) None of the above

201. Which of the following is true about Chemical simulation?
   a) Consider chemical reactions,
   b) Considers interaction of chemical additives with rock
   c) Both (a) and (b)
   d) None of the above

202. Which of the following methods is used to solve for linear system of equations (Finite)
   a) Explicit
   b) Implicit
   c) IMPES
   d) All of the above

203. Which of the following factors should be considered in reservoir simulation?
   a) Type of grid and dimension
   b) Type of reservoir and reservoir heterogeneity
   c) Number of fluid phases and components
   d) All of the above

204. Grid selection depends on:
   a) minimum spacing between wells
   b) Degree of reservoir heterogeneity
   c) Geometry of reservoir
   d) All of the above

205. What will be the skin factor resulting from the invasion of the drilling fluid to a radius of 2 feet from the information given below?

   Given Information:

   The permeability of the skin zone is estimated at 20 md as compared with the unaffected formation permeability of 60md. The wellbore radius is 0.25 ft.

   a) 4.16
   b) 3.16
   c) 2.16
   d) 5.16

206. An oil well is producing at a constant oil flow rate of 1200 STB/day under a semi steady-state flow regime. Well testing data indicate that the pressure is declining at a constant rate of 4.655 psi/hr. The following additional data is available: h =25 ft; Φ= 15%; Bo =1.3 bbl/STB; ct = 12 x10^{-6} psi^{-1}. The well drainage area will be:
   a) 1742400 ft^2
   b) 2742400 ft^2
   c) 3742400 ft^2
   d) 4742400 ft^2

Answers

Q1.	a	Q36.	a	Q71.	b	Q106.	a	Q141.	b
Q2.	b	Q37.	a	Q72.	d	Q107.	d	Q142.	c
Q3.	a	Q38.	d	Q73.	d	Q108.	c	Q143.	a
Q4.	a	Q39.	d	Q74.	d	Q109.	b	Q144.	a
Q5.	b	Q40.	d	Q75.	c	Q110.	c	Q145.	a
Q6.	d	Q41.	d	Q76.	d	Q111.	b	Q146.	a
Q7.	d	Q42.	a	Q77.	d	Q112.	b	Q147.	c
Q8.	a	Q43.	a	Q78.	c	Q113.	c	Q148.	a
Q9.	b	Q44.	d	Q79.	b	Q114.	c	Q149.	a
Q10.	b	Q45.	d	Q80.	c	Q115.	a	Q150.	a
Q11.	b	Q46.	d	Q81.	c	Q116.	a	Q151.	a
Q12.	d	Q47.	b	Q82.	a	Q117.	b	Q152.	a
Q13.	a	Q48.	a	Q83.	a	Q118.	d	Q153.	a
Q14.	c	Q49.	a	Q84.	b	Q119.	a	Q154.	a
Q15.	d	Q50.	a	Q85.	a	Q120.	a	Q155.	a
Q16.	d	Q51.	b	Q86.	a	Q121.	b	Q156.	a
Q17.	a	Q52.	a	Q87.	a	Q122.	a	Q157.	c
Q18.	a	Q53.	d	Q88.	d	Q123.	a	Q158.	a
Q19.	a	Q54.	c	Q89.	a	Q124.	d	Q159.	a
Q20.	d	Q55.	a	Q90.	c	Q125.	a	Q160.	a
Q21.	c	Q56.	a	Q91.	d	Q126.	b	Q161.	a
Q22.	c	Q57.	a	Q92.	c	Q127.	a	Q162.	c
Q23.	a	Q58.	d	Q93.	b	Q128.	b	Q163.	a
Q24.	b	Q59.	a	Q94.	a	Q129.	a	Q164.	a
Q25.	d	Q60.	a	Q95.	c	Q130.	a	Q165.	a
Q26.	a	Q61.	d	Q96.	b	Q131.	a	Q166.	b
Q27.	b	Q62.	a	Q97.	d	Q132.	a	Q167.	d
Q28.	b	Q63.	a	Q98.	d	Q133.	a	Q168.	a
Q29.	a	Q64.	c	Q99.	a	Q134.	a	Q169.	c
Q30.	c	Q65.	c	Q100.	a	Q135.	a	Q170.	b
Q31.	b	Q66.	d	Q101.	c	Q136.	a	Q171.	a
Q32.	c	Q67.	a	Q102.	c	Q137.	a	Q172.	b
Q33.	a	Q68.	a	Q103.	c	Q138.	a	Q173.	a
Q34.	c	Q69.	d	Q104.	c	Q139.	a	Q174.	b
Q35.	c	Q70.	d	Q105.	b	Q140.	a	Q175.	d

Answers

Q176.	b	Q183.	a	Q190.	d	Q197.	d	Q204.	d
Q177.	a	Q184.	a	Q191.	a	Q198.	d	Q205.	a
Q178.	b	Q185.	b	Q192.	b	Q199.	d	Q206.	a
Q179.	a	Q186.	a	Q193.	c	Q200.	c		
Q180.	b	Q187.	b	Q194.	d	Q201.	d		
Q181.	c	Q188.	c	Q195.	a	Q202.	d		
Q182.	d	Q189.	b	Q196.	d	Q203.	d		

CHAPTER 4

Petroleum Production Operations-I

1. The process of making a well ready for production (or injection) is called
   a) Well completion
   b) Workover
   c) Well stimulation
   d) None of the above

2. Well completion depends on
   a) Well production strategy
   b) Interface between wellbore and reservoir
   c) Number of Tubing strings,
   d) All of the above

3. Which of the completion has less degree of control over desired productive interval?
   a) Open hole completion
   b) Cased hole completion
   c) Both (a) and (b)
   d) None of the above

4. Which completion cannot be completed in layered formation of separate reservoirs with incompatible fluid properties?
   a) Open hole completions
   b) Perforated casing completions
   c) Both (a) and (b)
   d) None of the above

5. Which of the following is most suitable for a low pressure reservoir with unconsolidated heavy oil sand?
   a) Open hole completion
   b) Perforated casing completion
   c) Uncemented liner completion
   d) None of the above

6. Which of the following advantages are associated with uncemented liner completion?
   a) Less formation damage
   b) Log interpretation is less complicated
   c) Deepening the well can easily be carried out
   d) All of the above

7. Which of the following is true about packer?
   a) It provides a seal between the tubing and the tubing casing annulus
   b) It protect casing as well as prevent excessive tubing movement
   c) It isolates the producing intervals
   d) All of the above

8. Which of the following statement is true about the advantages of perforated casing completion?
   a) Gas or water production can be controlled easily
   b) Can be selectively stimulated
   c) Adaptable to multiple completion techniques
   d) All of the above

9. What are the limitations of Single String - Multiple packers multiple-zone completion?
   a) Monitoring flow from individual zones is difficult
   b) Difficulty of treating or re-perforating individual zone unless well is killed and tubing is pulled
   c) Both (a) and (b)
   d) None of the above

10. Which of the following is true about horizontal and multilateral well completion?
   a) These completions can be suitable in improving the productivity from the thin reservoirs.
   b) These completions improve recovery in tight low permeability zones
   c) Both (a) and (b)
   d) None of the above

11. Workover operation is more difficult in________
   a) Single zone completion
   b) Multiple zone completion
   c) Both (a) and (b)
   d) None of the above

12. Which of the following is true about monobore completion?
   a) It can be used for smaller and marginal field development
   b) It is difficult to achieve a good cementing job
   c) It can lead to corrosion of casing due to fluids being exposed to the casing
   d) All of the above

13. Which type of multilateral well completion is called open hole sidetrack?
   a) Level 1
   b) Level 2
   c) Level 3
   d) None of the above

14. In which type of multilateral well completion, both the main well and the secondary wells are open hole?
   a) Level 1
   b) Level 2
   c) Level 3
   d) None of the above

15. Which type of multilateral well completion has cased and cemented main well while the secondary wells have an openhole completion?
   a) Level 1
   b) Level 2
   c) Level 3
   d) None of the above

16. In which type of multilateral well completion, lateral liner is anchored inside the main well?

a) Level 1

b) Level 2

c) Level 3

d) None of the above

17. In which type of multilateral well completion, both the lateral well is cemented?

a) Level 1

b) Level 2

c) Level 3

d) Level 4

18. Which type of multilateral well completion uses three conventional packers?

a) Level 3

b) Level 4

c) Level 5

d) None of the above

19. Which type of multilateral well completion aims to guarantee mechanical and hydraulic continuity by using casing to ensure hydraulic sealing in the branching section?

a) Level 1

b) Level 2

c) Level 6

d) None of the above

20. What is the name for the wellhead and valve assembly used to control a well?

a) BOP

b) Annular preventer

c) Christmas tree

d) None of the above

21. What are the functions of the x-mas tree?

a) Regulate flow of hydrocarbons from the well by means of valves

b) Control gas or water injection into a well

c) Means of well intervention and Well monitoring: pressure, temperature, flow rate, sand detection

d) All of the above

22. ________ is used to connect flanges of different working pressure

a) Crossover flange

b) Adapter

c) Tool joint

d) None of the above

23. ________ is used to connect flanges of different dimensions or connect a flange to threaded end

a) Crossover flange

b) Adapter

c) Tool joint

d) None of the above

24. Which of the following is flow control equipment?

a) Surface & Subsurface Chokes

b) No-go nipples

c) Landing Nipples

d) All of the above

25. What is the name of the process by which communication is gained with the casing, cement and formation?
    a) Perforating
    b) Drilling
    c) Workover
    d) None of the above

26. Which of the following is required for optimum perforating gun systems?
    a) Sufficient downhole penetration to shoot well beyond any formation damage
    b) Phasing and maximum shot density
    c) Both (a) and (b)
    d) None of the above

27. Which of the following term is used to categories the firing of explosives?
    a) Order
    b) Phasing
    c) Shot density
    d) None of the above

28. Which explosives are used to initiate or detonate the explosive sequence?
    a) Primary Explosives
    b) Secondary Explosives
    c) Both of these
    d) None of the above

29. Which explosives are harder to initiate and less sensitive?
    a) Primary Explosives
    b) Secondary Explosives
    c) Both of these
    d) None of the above

30. Which of the following are high order explosives?
    a) Primary Explosives
    b) Secondary Explosives
    c) Both of these
    d) None of the above

31. Karakas and Tariq have presented a semi analytical solution for the calculation of the perforation skin effect. What are the different components of perforation skin effect:
    a) Plane flow effect, vertical converging effect and gravity effect
    b) Plane flow effect, vertical converging effect and well bore effect
    c) Well bore effect, vertical converging effect and gravity effect
    d) None of the above

32. A common way to attempt to decrease initial perforation damage is to:
    a) Perforate overbalanced
    b) Perforate underbalanced
    c) Perforate as close to balanced as possible
    d) None of the above

33. Which of the following is the example of primary high explosives?
    a) Zinc azide
    b) TNT

c) RDX

d) None of the above

34. Which of the following is the example of secondary high explosives?

a) Zinc azide

b) Lead styphhnate

c) RDX

d) None of the above

35. The liner geometry of Deep penetrating shape charge is

a) Conical

b) Parabolic

c) Hemispherical

d) None of the above

36. The liner geometry of Big hole shape charge is

a) Conical

b) Parabolic

c) Square

d) None of the above

Statement for Linked Answer Questions 37-38

The following information are given coiled tubing conveyed perforating job:

Reservoir Depth: 8000 ft TVDSS

Reservoir Pressure: 1660 psi

Desired perforating underbalance : 500 psi

Desired CT pressure for TCP detonation: 500 psi

Well bore fluid: 8.5 ppg brine

Use 20% safety factor for hydraulic disconnect

37. The hydraulic firing head will detonate at an absolute pressure of :

a) 4035 psi

b) 5035 psi

c) 6035 psi

d) 3035 psi

38. The pressure required to the coil to the coiled tubing to shear release the hydraulic disconnect is

a) 5035 psi

b) 2450 psi

c) 4000 psi

d) 3000 psi

39. What type of down hole tool is needed to complete a high pressure squeeze cementing job?

a) Packer.

b) Liner

c) Casing

d) Tubing

40. When a producing well has to be shut in for remediation operations it is called?

a) Well stimulation

b) Workover

c) Both of these

d) None of the above

41. What is needed for coiled tubing operations that are not needed for normal drilling rigs?

a) BOP

b) Wellhead

c) Injector

d) None of the above

42. The process of creating holes in the casing that pass through the cement sheath and extend some depth into the formation
    a) Perforation
    b) Fracturing
    c) Acidizing
    d) None of the above

43. What is the primary method of stimulation for sandstone formations?
    a) Matrix Acidizing
    b) Acid Fracturing
    c) Hydraulic Fracturing
    d) None of the above

44. What is the primary method of stimulation for carbonate formations?
    a) Matrix Acidizing
    b) Acid Fracturing
    c) Hydraulic Fracturing
    d) All of these

45. Which of the following acid is used in the main flush of matrix acidizing for sandstones
    a) HCl
    b) 3wt%HFand 12wt%HCl
    c) H_2SO_4
    d) None of the above

46. Which of the following acid is used in the main flush of matrix acidizing for carbonates
    a) HCl
    b) 3wt%HFand 12wt%HCl
    c) H_2SO_4
    d) None of the above

47. Which of the following is not a key assumption for matrix acidizing?
    a) Acid penetrates uniformly and radially
    b) Formation is homogeneous
    c) Pore size varies widely
    d) Reaction rate declines uniformly with decreasing acid concentration

48. Which of the following operation does not require sand?
    a) Hydrajet perforation
    b) Acid Fracturing
    c) Hydraulic Fracturing
    d) None of the above

49. Which of the following is the main objective of hydraulic fracturing?
    a) To place a conductive channel through the near well bore damage and thus bypass this crucial zone
    b) To extend a channel to a significant depth into the reservoir to increase the productivity
    c) Both (a) and (b)
    d) None of the above

50. Which of the following is good candidate for hydraulic fracturing?
    a) Low permeability reservoir
    b) High permeability reservoir
    c) Low near well bore damage
    d) None of the above

51. The ratio of the ability of a fracture to carry flow to the ability of the formation to feed the fracture is called:

 a) Dimensionless fracture conductivity
 b) Dimensionless fracture gradient
 c) Both (a) and (b)
 d) None of the above

52. The minimum in situ stress can be determined by

 a) Minifrac test
 b) Drillstem test
 c) Reservoir limit test
 d) None of the above

53. The fracturing fluid coefficient is a key factor in modeling an induced fracture. Which of the following is not a category for fracture fluids?

 a) Conductivity-controlled fluids
 b) Reservoir-controlled fluids
 c) Viscosity-controlled fluids
 d) Wall-building fluids

54. PKN fracture model predicts

 a) short and wider fracture
 b) longer and thinner fracture
 c) both (a) and (b)
 d) None of the above

55. KGD models predicts

 a) short(length) and wider fracture
 b) longer(length) and thinner fracture
 c) both (a) and (b)
 d) None of the above

56. Which of the following is true about the PKN model?

 a) It has an elliptical shape at the wellbore
 b) The maximum width is at the centerline of the ellipse
 c) Both (a) and (b)
 d) None of the above

57. Which of the following situations is the most ideal case for matrix simulation?

 a) A sandstone reservoir with low permeability and low formation skin damage
 b) A carbonate reservoir with high permeability and high formation skin damage
 c) A sandstone reservoir with high permeability and high formation skin damage
 d) A carbonate reservoir with low permeability and low formation skin damage

58. Following information are given for hydraulic fracturing job:

 Injection rate = 40 bpm

 Apparent viscosity of fracturing fluid = 100 cP

 Fracture half length = 1000 ft

 Youngs Modulus = 4×10^6 psi

 Poisson Ratio = 0.25

 Using PKN model, what would be the maximum and average fracture width

 a) 0.42 and 0.25
 b) 0.35 and 0.20
 c) 0.60 and 0.40
 d) None of the above

Statement for Linked Answer Questions 59-61

It is planned to carry out a matrix acidizing job for carbonate formations having following characteristics:

Formation Depth = 7500 ft,

Perforated Interval = 20 ft,

Formation Permeability = 10 md,

Fracture gradient at initial pressure of 3075 psi = 0.7 psi/ft

Overburden gradient = 1.0 psi/ft

Current reservoir pressure = 2000 psi

Acid viscosity at reservoir temperature = 0.4 cp

Acid hydrostatic gradient = 0.43 psi/ft

Drainage radius = 660 ft

Wellbore radius = 0.25 ft

59. The fracture gradient at current reservoir pressure would be
    a) 0.5 psi/ft
    b) 0.6 psi/ft
    c) 0.7 psi/ft
    d) 0.8 psi/ft

60. The maximum injection rate(in practice) would be
    a) 0.66 psi/ft
    b) 0.77 bbl/min
    c) 0.66 psi/ft
    d) 0.8 psi/ft
    e) None of the above

61. The maximum surface injection pressure would be:
    a) 1500 psi
    b) 2000 psi
    c) 3000 psi
    d) None of the above

Statement for Linked Answer Questions 62-64

It is planned to carry out a matrix acidizing job for sand stone formations of an oil well having following characteristics:

Formation Depth = 5000 ft,

Perforated Interval = 10 ft,

Formation Permeability = 50 md,

Fracture gradient at initial pressure of 2000 psi = 0.7 psi/ft

Overburden gradient = 1.0 psi/ft

Current reservoir pressure = 1000 psi

Acid viscosity at reservoir temperature = 0.75 cp

Acid hydrostatic gradient = 0.43 psi/ft

Drainage radius = 660 ft

Wellbore radius = 3 in.

62. The fracture gradient at current reservoir pressure would be
    a) 0.5 psi/ft
    b) 0.6 psi/ft
    c) 0.7 psi/ft
    d) 0.8 psi/ft

63. The maximum injection rate(in practice) would be
    a) 0.66 bbl/min
    b) 0.77 bbl/min
    c) 0.66 psi/ft
    d) 0.8 psi/ft
    e) None of the above

64. The maximum surface injection pressure would be:
    a) 1500 psi
    b) 2000 psi
    c) 3000 psi
    d) None of the above

65. A recently-drilled gas well has been completed, but initial flow rates from the well are disappointing. The well has the following characteristics:

The pay zone perforations are centered at 15,800 ft (4,816 m).

The static reservoir pressure is 0.63 psi/ft (1.45 SG).

The fracture gradient at the perforations is 0.95 psi/ft (2.19 SG).

Skin = -20 psi (-138 KPa).

The pay zone is a tight, highly-consolidated clean sandstone. Permeability is estimated to be 0.05 mD.

What treatment technique would you recommend to stimulate this well?

a) High rate, high volume fracture stimulation

b) Low rate "matrix" hydrochloric acid stimulation

c) Sand control treatment to reduce sand collection in the wellbore

d) Scale removal workover

66. Which of the following is used as breaker for fracturing fluids?

a) Ammonium persulphate

b) Ammonium sulphate

c) Diammonium peroxidisulphate

d) All of the above

67. The closure stress for sand and resin coated proppants is

a) <6,000 psi

b) 5,000-10,000 psi

c) >10,000 psi

d) None of the above

68. The closure stress for ceramic proppant is

a) <6,000 psi

b) 5,000-10,000 psi

c) >10,000 psi

d) None of the above

69. The closure stress for Sintered bauxite proppant is

a) <6,000 psi

b) 5,000-10,000 psi

c) >10,000 psi

d) None of the above

70. Surface chokes are usually sized for ___ flow

a) Critical

b) sub-critical

c) both (a) and (b)

d) None of the above

71. Subsurface chokes are usually sized for_____ flow

a) Sub-critical

b) Critical

c) Both of these

d) None of the above

72. No-Go Landing Nipples are used for

a) to prevent the passage of larger diameter wireline tools and control devices

b) to prevent the passage of smaller diameter wireline tools and control devices

c) Making the passage of large diameter wireline tools and control devices

d) None of the above

73. No-Go landing nipples installed at

a) at the bottom of the tubing below any selective nipple

b) in a tail pipe below a permanent packer

c) both (a) and (b)

d) None of the above

74. Which of the following is a circulating device

a) Sliding sleeve

b) Ported collars

c) Ported nipples

d) All of the above

75. The device that is run to form a seal between the tubing and casing

a) Packer

b) Sliding sleeve

c) Ported nipple

d) Ported collar

76. The function of blast joints

a) Protect the tubing against erosion from the jetting action of perforations

b) Used to prevent the passage of larger diameter wireline tools and control devices

c) Used to shut-in the well automatically in an emergency

d) All of the above

77. The device that is run to separate two production packers is

a) Sliding sleeve

b) Ported nipple

c) Ported collar

d) All of the above

78. Which of the following is true about flow coupling?

a) Protect the tubing from flow turbulence

b) Desirable where there is a flow restriction

c) Both (a) and (b)

d) None of the above

79. Which of the following is true about blast joints?

a) Made of extra-thick, heat treated, high strength steel alloys

b) Protect the tubing against erosion from the jetting actions of perforations

c) Same ID as tubing

d) All of the above

80. Which of the following is the advantage of surface controlled subsurface safety valve over subsurface controlled subsurface safety valve?

a) Can run wireline tools through valve

b) Less affected by sand production

c) Easily and Economically retrieved for inspection and repair

d) All of the above

81. The standard techniques of conveying perforating guns include

a) Through-Casing perforation

b) Through-Tubing perforation(TTP)

c) Tubing Conveyed perforation (TCP)

d) All of the above

82. Which of the following is an important parameter for the design of Gravel Packing system?

a) Optimum gravel size to formation sand size

b) Optimum screen slot width to retain gravel

c) Effective placement technique

d) All of the above

83. Which of the following component is the basic element of the packers?

a) Cones

b) Slips

c) Packing Materials

d) All of the above

84. Which of the following is true about Tension-set packers?

a) the change in pressure from below pushes on tubing and sets packer harder

b) Used on shallow applications.

c) Set by rotate and pull tubing

d) All of the above

85. Which of the following is true about Compression set packer?

a) high pressure change below sets slips harder and high pressure applications (hydraulic hold downs).

b) Used for injection wells, on high BHP and frac/ acid.

c) Set by rotate and slackoff.

d) All of the above

86. Which of the following is true about Inflatable production packer?

a) steel reinforced strips stamped with special tabs, which grip into casing or open hole when inflated

b) Can be retrievable packers, temporary or permanent bridge plugs, or liner hangers.

c) Both (a) and (b)

d) None of the above

87. Which of the following is true about Swabbing?

a) Pulling fluid from the well bore through the use of wire rope and cup assembly.

b) Swabbing equipment includes a swabbing assembly, lubricator with an oil saver, and shut-off valve on the well, also called a swabbing valve.

c) Both (a) and (b)

d) None of the above

88. Which of the following is true about Snubbing?

a) the control of a tubing string while running it in or out of a well bore under pressure

b) Pulling fluid from the well bore through the use of wire rope and cup assembly.

c) Both (a) and (b)

d) None of the above

89. Which of the following is true about Bullheading?

a) It involves pumping the kill fluid directly down the well bore, forcing the well bore fluids back into the reservoir.

b) Effective to well kill; building up a sufficient hydrostatic head in the well bore.

c) It can risk damaging the reservoir, by forcing undesired materials into it.

d) All of the above

90. The control pressure for a surface safety valve is most likely equal to:

a) Valve body pressure divided by actuator ratio

b) Actuator ratio divided by valve body pressure

c) Two times valve body pressure divided by actuator ratio

d) Valve body pressure multiplied by actuator ratio

91. Tubing elongation most likely occurs with:

a) Ballooning

b) Buckling

c) Reverse Ballooning

d) Temperature Reduction

92. Which of the following is not a typical class of pipe cutters?

a) Acid

b) Jet-shaped charge

c) Outside cutter

d) Tubing punch

93. A caliper survey has been run on a producing gas well. It shows that the casing is partially collapsed 500 ft (152.4 m) above the perforations. The well is starting to load up with fluid, so a decision has been made to install a small diameter tubing string with a packer set just below the partially collapsed interval to protect the casing from pressure, temperature fluctuations and other forces that could result in further collapse. What type of packer should be used in this application?

a) Weight-set packer

b) Tension-set packer

c) Rotation-set packer

d) Hydraulic-set packer

94. The following informations are given for fracing job:

Modulus of elasticity of steel: 30 × 106psi

Coefficient of thermal expansion for steel: 6.9×10–6 in/in/°F

Poisson's ratio: 0.30

Packer set at: 8,500 feet

Tubing I.D. 2.992 inches; Area 7.031 in2 (frac string)

Tubing O.D. 3.5 inches; Area 9.621 in2

Casing I.D. 6.094 inches; Area 29.17 in2

Casing O.D. 7.000 inches; Area 38.49 in2

Seal bore packer:

Seal bore I.D. 3.00 inches Area 7.07 in2

Seal I.D. 2.28 inches Area 4.08 in2

Initial conditions:

Surface casing pressure: 0 psi

Surface tubing pressure: 0 psi

Surface temperature: 75°F

Bottomhole temperature: 160°F

Density of fluid (both tubing and annulus): 9.5 lb/gal

Final pumping conditions:

Surface casing pressure: 500 psi

Surface tubing pressure: 5,500 psi

Bottomhole tubing pressure (@ packer): 7,000 psi

Surface temperature: 75°F

Bottomhole temperature: 75°F

Density of fluid in annulus: 9.5 lb/gal

If the tubing is free to move, the length change (inches) due to the piston effect is most nearly:

a) –8.2

b) -1.8

c) -1.4

d) +0.1

95. It is planned to perforate a sand at 9,160 ft with a 500-psi underbalance. The sand has a 4,525-psi reservoir pressure, and the hole is full of 9.5-lb/gal fluid. Assuming any nitrogen cushion would have a gradient of 0.1 psi/ft, the following condition that would achieve the desired underbalance is most nearly:

a) Existing well fluid swabbed to 3,050 ft plus a nitrogen cushion with a 700-psi surface pressure

b) Existing well fluid swabbed to 2,030 ft plus a nitrogen cushion with an 800-psi surface pressure

c) Existing well fluid swabbed to 2,630 ft plus a nitrogen cushion with an 800-psi surface pressure

d) All of the above are within 25 psi of the desired underbalance

96. A mini-frac was performed using 8.34-lb/gal frac fluid at a pump rate of 30 BPM. The treatment interval had an initial reservoir pressure of 3,000 psi and a depth of 6,000 ft. The following wellhead pressures were obtained during the mini-frac:

Pre-job pressure 402 psi

Average treatment pressure 2,110 psi

Maximum treating pressure 2,238 psi

Initial shut-in pressure 1,710 psi

The frac gradient (psi/ft) is most nearly:

a) 0.50

b) 0.72

c) 0.78

d) 0.81

97. The reservoir capability can be best illustrated by

a) Inflow Performance Relationship Curve

b) Tubing Intake Curves

c) Wellhead Performance Relationship curve

d) None of the above

98. The piping capability can be best illustrated by
   a) Inflow Performance Relationship Curve
   b) Tubing Intake Curves
   c) Wellhead Performance Relationship curve
   d) None of the above

99. What is the unit of the productivity index?
   a) STB/PSI
   b) (STB/D)/PSI2
   c) PSI/(STB/D)
   d) (STB/D)/PSI

100. Productivity Index can be mathematically expressed as
   a) Production X Drawdown
   b) Production / Drawdown
   c) Production + Drawdown
   d) Production – Drawdown

101. The AOF is the:
   a) Minimum gas rate of the tested well
   b) Extent of the drainage area
   c) Maximum theoretical gas rate with $P_{wf} = P_{atm}$
   d) Gas well productivity index

102. Which of the following is not a component of IPR?
   a) Reservoir pressure
   b) Perforation radius
   c) Completion thickness
   d) Fluid density

103. An IPR curve will not be a straight line if
   a) The only produced fluid is gas
   b) The only produced fluid is oil
   c) The only produced fluid is water
   d) All IPR plots are curved

104. An IPR curve can be a straight line if:
   a) The only produced fluid is gas with condensate
   b) The only produced fluid is oil
   c) Both oil and gas are produced
   d) Pwf is at least 100 psi below the bubble point

105. Which of the following is true about homogeneous flow TPR models?
   a) The multiphases are treated as a homogeneous mixture
   b) These models consider the effects of liquid hold up
   c) Both (a) and (b)
   d) None of the above

106. Which of the following is true about separated flow models?
   a) These flow models consider the effect of liquid hold up
   b) These flow models consider flow regime
   c) Both (a) and (b)
   d) None of the above

107. A gas reservoir is being produced at a rate of 500 MSCF/d. The pressure measurement indicated much higher bottom-hole flowing pressure than calculated based on Darcy law IPR. What could be a possible reason for this higher bottom-hole flowing pressure?

a) Non-Darcy flow

b) Lower skin

c) Lower permeability

d) None of the above

108. If pressure drop in the wellbore is reduced while all other parameters are the same, that will result in:

a) Increased bottom-hole pressure,

b) Reduced wellhead tubing flow pressure,

c) Increased production rate,

d) Change in IPR curve

109. A reservoir is producing at a production rate that is half of its potential. To determine whether formation damage, or flow assurance in the wellbore is causing the problem a nodal analysis is conducted. Where would you select as a node to determine which is causing the problem?

a) At the wellhead

b) At the bottom-hole near the completion

c) At the separator

d) At midpoint in wellbore

110. Which of the following is the example of homogeneous flow model?

a) Poetman and Carpenter Correlation

b) Lockhart and Martinelli Correlation

c) Duns and Ros Correlation

d) None of the above

111. Which of the following is the example of separated flow model?

a) Poetman and Carpenter Correlation

b) Guo and Ghalambor correlation

c) Hagedorn and brown correlation

d) None of the above

112. In analyzing performance of a production system, two pressures are considered constant. They are:

a) Pbottomhole and Pseparator

b) Preservoir and Pbottomhole

c) Pbottomhole and Pchoke

d) Preservoir and Pseparator

113. The actual production rate of a reservoir is lower than expected based on reservoir analysis. What can be a possible cause of this lower production?

a) Higher formation permeability

b) Higher reservoir pressure

c) Higher skin

d) All of the above

114. Which flow pattern is expected if the entire zone of interest is perforated?

a) Radial
b) Linear
c) Spherical
d) Both A and B are possible

115. Which of the following can cause the IPR curve to become a straight line?

a) Skin equal to zero
b) Transient flow
c) bottom-hole pressure below the bubble point
d) Single-phase oil flow
e) None of the above

116. One of the followings is true about turbulence

a) turbulence decreases pressure loss
b) Turbulence increases pressure loss
c) Turbulence occurs when the flow becomes stable
d) All of the above

117. In ________gas is dispersed in the form of small bubbles in a continuous phase

a) Bubble flow
b) Slug flow
c) Churn flow
d) None of the above

118. In ________ gas bubbles coalesce into larger bubbles that are eventually fill the entire pipe cross section

a) Slug flow
b) Churn Flow
c) Bubble flow
d) Annular flow

119. In ________ flow the larger gas bubbles become unstable and collapse, resulting in a highly turbulent flow pattern with both phases dispersed

a) Slug flow
b) Churn Flow
c) Bubble flow
d) Annular flow

120. A production engineer can do several things to increase production from an oil reservoir which is producing at steady-state condition. Which is one method that a production engineer can use to increase production from the above reservoir?

a) Decrease wellbore size
b) Increase bottom-hole flowing pressure
c) Increase reservoir pressure
d) None of the above

121. Which parameter has no effect on the inflow performance of a reservoir?

a) Formation permeability
b) Perforation skin
c) Production tubing diameter
d) Reservoir boundary condition
e) PVT behavior

122. An IPR curve for a natural gas reservoir has what shape?

a) Straight line

b) Straight line that curves at low bottom-hole pressure

c) Curved line through all bottom-hole pressures

d) Curved line at high bottom hole pressure, straight at low pressure

123. A production engineer can enhance production of a well by:

a) Increasing bottom-hole flowing pressure

b) Increasing surface flowing pressure

c) Increasing reservoir pressure

d) None of the above

124. In _______ , gas bubbles becomes the continuous phase with liquid flow in the annulus coating the surface of the pipe and with droplets entrained in the gas phase.

a) Bubble flow

b) Slug flow

c) Churn flow

d) Annular flow

125. The better suited Inflow Performance Relationship for wells flowing below bubble point is

a) Straight Line Relationship

b) Vogels IPR

c) None of the above

d) Orkiszewski 's IPR

126. Maximum quantity of oil can be recovered in the case of

a) Active water drive reservoirs

b) Depletion/Solution gas reservoirs

c) Gas cap driven reservoirs

d) Data insufficient

127. If the bottomhole flowing pressure is reduced due to some changes in the reservoir properties with the same production rate as before, then the new productivity index of the well is compared to the previous productivity index of the well is:

a) Higher

b) Lower

c) The same

d) Cannot be determined from the give information

128. If the actual production rate of a well is lower than that predicted based on IPR equation, and the bottomhole pressure measurement shows it to be lower than expected, then the cause is:

a) By a restriction in the wellbore

b) By formation damage

c) By an increase in reservoir pressure

d) None of the above

129. In oil well design, prediction of multiphase flow pressure drop is helpful for

a) Deciding on tubing size,

b) Deciding on flow line size,

c) Both of the above

d) None of the above

130. Out of the following, which pressure is the highest in a flowing well
   a) Well head Tubing pressure (WHP)
   b) Flowing Bottom hole pressure (FBHP)
   c) Static bottom hole pressure (SBHP)
   d) None of the above

131. Draw down is defined as
   a) Difference between reservoir & static pressure
   b) Difference between reservoir and flowing bottom hole pressure
   c) Difference between flowing bottom hole pressure and tubing head pressure
   d) Difference between reservoir and tubing head pressure

132. Which of the following parameters does not affect tubing performance or VLP?
   a) Tubing wellhead flowing pressure
   b) Perforations
   c) Reservoir fluid properties
   d) Both (b) and (c)

133. Intersection point of IPR and TIC curves indicates
   a) Possible flow rate from the well
   b) Quantity of formation damage in the well
   c) None of the above
   d) Reservoir pressure

134. Solution gas drive is also called
   a) Segregation drive
   b) Depletion drive
   c) Gas drive
   d) None of the above

135. Undersized tubing will result in:
   a) Higher wellhead pressure than optimum
   b) Higher bottom-hole pressure than optimum
   c) Lower bottom-hole pressure than optimum
   d) Need more information

136. The purpose of artificial lift is
   a) To create required steady flowing bottom hole pressure
   b) To increase reservoir pressure
   c) To increase tubing head pressure
   d) None of the above

137. Which of the following has the least volumetric efficiency for rod pumped wells?
   a) High fluid level with no gas interference
   b) Gaseous wells with fair downhole separation
   c) Foamy liquid or gaseous conditions
   d) Ample pump submergence with good separation

138. The primary function of the counter-balance system in rod pumped wells is:

a) Speed reduction of the prime mover to suitable pumping speeds

b) Store energy on the down-stroke and release energy on the upstroke

c) Store energy on the up-stroke and release energy on the downstroke

d) Supply the mechanical energy to transmit to the pump and lift fluids

139. Which of the following is false about the API designation of a rod pump?

a) TP stand for tubing pump with heavy wall and a soft packed plunger

b) TP stand for tubing pump with heavy wall and a metal plunger

c) RWB pump is a rod pump with a thin wall stationary barrel, metal plunger and a bottom anchor

d) None of the above

140. Which type of SRP is better suited for producing sand laden well fluids

a) Top and bottom anchor rod pump

b) Ring valve pump

c) Both (a) and (b)

d) None of the above

141. Which of the following is true?

a) Moulded rod guide systems are used for inclined wells

b) Three tube pump is ideally suited to produce sand laden fluids

c) Both (a) and (b)

d) None of the above

142. The operational change that would least reduce the torque experienced by a pumping unit system is most likely:

a) Reducing the stroke length

b) Reducing the unit's speed

c) Changing the direction of rotation

d) Installing a larger rod string

143. In SRP operation, gas interference

a) Increases pumping efficiency

b) Decreases pumping efficiency

c) Does not affect pumping efficiency

d) None of the above

144. In a C-456D-256-144 SRP pumping unit , 144 means

a) 144 inches of max possible stroke length

b) 144000 inch-lb of max gear-box torque

c) 14400 lb Peak polish rod load capacity

145. Very high viscous crude can efficiently be lifted by
   a) Gas lift
   b) Sucker Rod Pump
   c) Progressive cavity pump.
   d) None of the above

146. Dynagraph is a record of
   a) Rod load and rod position
   b) Rod torque and power
   c) Rod torque and rod position
   d) None of the above

147. In a C-456-256-144 unit, 256 means
   a) gear box having 256000 inch-lb torque maximum
   b) Peak polished rod load capacity of 25600 lbs
   c) Stroke length of 256 inches.

148. Which of the following is true about the gas lift?
   a) Gas lift is a mechanical process of lifting fluid from a well bore
   b) Gas lift is a chemical process of lifting the fluid from well bore
   c) Gas lift is a electromagnetic process of lifting the fluid from the well bore
   d) None of the above

149. As per gas lift applications, what are the high productivity index high bottom hole pressure wells?
   a) The wells, which have productivity index greater than 0.5 and bottom hole pressure of the well supports a fluid column equal to 70% of the well depth are called high productivity index high bottom hole pressure wells
   b) The wells, which have productivity index greater than 0.5 and bottom hole pressure of the well supports a fluid column equal to 60% of the well depth are called high productivity index high bottom hole pressure wells.
   c) The wells, which have productivity index greater than 0.5 and bottom hole pressure of the well supports a fluid column equal to 50% of the well depth are called high productivity index high bottom hole pressure wells.
   d) None of the above

150. As per gas lift applications, what are the high productivity index low bottom hole pressure wells?
   a) The wells, which have productivity index greater than 0.5 and bottom hole pressure of the well supports a fluid column less than 50 % of the well depth are called high productivity index low bottom hole pressure wells
   b) The wells, which have productivity index greater than 0.5 and bottom hole pressure of the well supports a fluid column less than 40 % of the well depth are called high productivity index low bottom hole pressure wells

c) The wells, which have productivity index greater than 0.25 and bottom hole pressure of the well supports a fluid column less than 60 % of the well depth are called high productivity index low bottom hole pressure wells

d) None of the above

151. Wells with a high bottom hole pressure and a high productivity index are normally designed for:

a) Semiclosed continuous flow

b) Semiclosed intermittent flow

c) Closed intermittent flow

d) None of the above

152. Wells with a high bottom hole pressure and low productivity index are normally designed for:

a) Semiclosed continuous flow

b) Semiclosed intermittent flow

c) Closed intermittent flow

d) None of the above

153. Wells with a low bottom hole pressure and either a high or low productivity index are normally designed for:

a) Semiclosed continuous flow

b) Semiclosed intermittent flow

c) Closed intermittent flow

d) None of the above

154. For the flow of 2000 bbl/d of oil and 1MMSCF/d of gas at a flowing tubing pressure of 800 psi, the diameter of choke using Gilbert correlation would be --- in 64th of an inch

a) 30

b) 32

c) 33

d) 35

155. Which of the following is the basic component of gas lift valve?

a) Metal bellow

b) Metal rod

c) Orifice or port

d) All of the above

156. For intermittent operations, it is generally recommended to use the:

a) valves of port size ½ in. or more

b) valves of port size less than ½ in.

c) Both (a) and (b)

d) None of the above

157. For continuous lift operations, it is generally recommended to use the:

a) valves of port size ½ in. or more

b) valves of port size less than ½ in.

c) Both (a) and (b)

d) None of the above

158. The basic characteristics of intermittent gas lift wells are:

a) Low BHP and Low PI or high PI

b) Low producing rate

c) Deep with low fluid level

d) All of the above

159. For bellows-type gas-lift valves, valve spread is:

a) The difference between opening and closing pressures

b) The difference between annular and bellow pressures

c) The difference between Casing pressure and tubing pressure against the valve

d) The difference between Reservoir pressure and flowing bottom hole pressure.

160. In test rack setting conditions, tubing pressure acting on the valve is

a) Maximum

b) Zero

c) Both (a) and (b)

d) None of the above

161. Gaslift Optimization curve is plotted with

a) Liquid production and well head pressure

b) Liquid production and Gas injected

c) Gas injected and Formation gas

d) None of the above

162. Point of balance in gas lift well means

a) Intersection of injection pressure and reservoir pressure

b) Intersection of tubing pressure and injection pressure

c) Both (a) and (b)

d) None of the above

163. Number of valves required in a gas lift well primarily depend upon

a) Well depth and Injection pressure

b) Well depth and Reservoir pressure

c) Well depth only

164. Intermittent wells are referred to as least efficient due to

a) High back pressure

b) High liquid fall back in tubing

c) High Injection pressure requirement

d) None of the above

165. ESP is basically suited for

a) Low volumes of well fluids

b) Sand laden well fluids

c) High volume of well fluids

d) None of the above

166. ESP belongs to the following class of mechanical pumps

a) Positive displacement type

b) Centrifugal pump

c) Both of the above

d) None of the above

167. In the case of ESP, increasing number of stages will help to

a) Increase the head developed

b) Increase the flow rate of the pump

c) Both of the above

d) None of the above

168. In ESP , the motor is generally
   a) Attached to the bottom-most part of the assembly
   b) Attached at the top-most part of the assembly
   c) Attached at the surface near well head
   d) None of the above.

169. Electric submersible pumps are
   a) Centrifugal pumps in multi stage fashion
   b) Reciprocating pumps
   c) Positive displacement pump
   d) None of the above

170. No of stages required by an ESP is governed by
   a) Head of the fluid to be lifted to surface against a given tubing pressure
   b) Fluid quantity required at the surface
   c) Both
   d) None of the above

171. Which of the following artificial lift technique is used for offshore wells?
   a) Sucker Rod Pumping System
   b) Gas Lift
   c) Electrical Submersible Pumping System
   d) Both (b) and (c)

172. Jet pump has
   a) No moving parts
   b) One moving part
   c) Some moving parts
   d) None of the above

173. One inch sucker rod weighs 2.88 lb per ft. The maximum polished rod load resulting from 2000 ft of one-inch sucker rods is ---, if the pumping speed is 18 spm and the polished rod stroke length is 74 in.
   a) 7120 lb
   b) 7720 lb
   c) 8250 lb
   d) 7250 lb

174. If the prime mover shaft rotates at 1000 rpm for a 20 spm pump, this speed must be reduced to ---of the prime movers speed by the time the power reaches the crankshaft.
   a) 1/50
   b) 1/5
   c) 1/10
   d) 1/20

175. A well is operating on intermittent gas lift. The injection gas pressure is 700 psi and valve depth is 4500 ft. The cycle frequency for intermittent lift operation would be
   a) 152 cycle per day
   b) 107 cycle per day
   c) 204 cycle per day
   d) 45 cycle per day

176. Very high viscous crude can be efficiently lifted by
   a) Jet pump
   b) Sucker Rod Pump
   c) Progressive cavity pump
   d) None of the above

177. A well produces 100 stb/day at a measure flowing bottom-hole pressure of 1000 psia. The average reservoir pressure is 2000 psia. The productivity index of the tested well in stb/day-psi will be:

a) 0.1 stb/day.psi

b) 0.2 stb/day. psi

c) 0.3 stb/day. psi

d) None of the above

178. The most critical component of a PCP is

a) Rotor

b) Motor

c) Sucker rods

d) Elastomer in the stator.

Statement for Linked Answer Questions 179-181

The reservoir and well test data of an oil well are given below:

Permeability, K_o = 30 md

Pay thickness, h = 40 ft

Average reservoir pressure, p_r = 3000 psig

Reservoir temperature, T = 200°F

Well spacing, A = 160 acres (43560 ft²/acre)

Drilled hole size, D = 12-1/4 inch (open hole)

Formation volume factor, B_o = 1.2 (bbl/stb)

Oil viscosity, μ_o = 0.8 cp

(Assume skin = 0 and no turbulence)

179. The absolute open flow potential for the above well is:

a) 3672 stb/day

b) 4200 stb/day

c) 5000 stb/day

d) 3100 stb/day

180. The Productivity Index for the above mentioned oil-well is:

a) 1.8 stb/psi-day

b) 1.22 stb/psi-day

c) 0.5 stb/psi-day

d) 2.5 stb/psi-day

181. The production rate for the above oil well at p_{wf} = 1000 psig is:

a) 2600 stb/day

b) 2440 stb/day

c) 2800 stb/day

d) 2920 stb/day

182. A 1000 ft-long horizontal well is drilled in a reservoir with the following characteristics:

k_h = 75 md

B = 1.34 res.bbl/st.bbl

h = 30 ft

μ = 0.62 cp

r_{eh} = 1053 ft

The productivity index for this horizontal well is:

a) 13.3 stb/psi-day

b) 21.0 stb/psi-day

c) 18.0 stb/psi-day

d) 25.0 stb/psi-day

183. A buildup test in a well after a constant rate production, q_o=100BPD indicates Kh/μ = 20 md-ft/cp, S_t = 2. The pressure loss due to skin for B_o =1 is:

a) 1520 psi

b) 1350 psi

c) 1412 psi

d) 1380 psi

184. The following information are given for an oil well:

K=5md

p_r = 2500 psig

h = 20ft

s = -1

μ_o = 1.1 cp

B_o = 1.2 RB/STB

Spacing = 80 acres

r_w = 0.365 ft

The absolute open flow potential for this well is:

a) 290 stb/day

b) 216 stb/day

c) 250 stb/day

d) 200 stb/day

185. A well is producing from a saturated reservoir with an average reservoir pressure of 2500 psig. Stabilized production test data indicated that the stabilized rate and wellbore pressure are 350 STB/day and 2000 psig, respectively.

The oil production rate at flowing bottom hole pressure of 1850 psi assuming constant PI would be :

a) 1067.1 STB/day

b) 967.1 STB/day

c) 1167 STB/day

d) 1267 STB/day

186. A well is producing from a saturated reservoir with an average reservoir pressure of 3000 psig. Stabilized production test data indicated that the stabilized rate and wellbore pressure are 500 STB/day and 2500 psig, respectively. The oil production rate at flowing bottom hole pressure of 1200 psi using vogel's equation will be

a) 2415 stb/day

b) 1415 stb/day

c) 3415 stb/day

d) None of the above

187. An oil well is producing under a constant bottom hole flowing pressure of 1500 psi. The current reservoir pressure is 3200 psi. The well is developed in the center of a 40 acre square drilling pattern. Giving the following additional information

Φ = 16%; h=15 ft; k =50 md; μ = 26 cp; Bo =1.15 bbl/STB, C = 10 x 10^{-6} psi^{-1}

The flow rate of the oil well will be:

a) 416 STB/day

b) 316 STB/day

c) 516 STB/day

d) 216STB/day

188. The choke size that will produce the well at a rate of 415 STB/D for the given data using Gilbert choke equation will be:

Given data:

F_{gL} = 600

B = 0.546

p_1 = 200 psia

C = 1.89

A = 10.0

D = 14.7

q_L = 415 STB/D

a) 48 (1/64th of an inch)

b) 32 (1/64th of an inch)

c) 24 (1/64th of an inch)

d) None of the above

189. A 24-hour production test was obtained for a new oil well completed with 2 7/8-inch-O.D., 6.5-lb/ft tubing. The following well and test information is available for analysis:

Well data:

Well depth 6,502 ft

Casing setting depth 6,492 ft

Tubing setting depth 6,003 ft

Perforations (mid-perf) 6,076 ft

Reservoir data:

Initial reservoir pressure 2,800 psia

Bubble point pressure 2,400 psia

Current reservoir pressure 2,200 psia

Production test data:

Oil production rate 421 bbl/d

Gas production rate 170 Mscf/d

Flowing bottom hole pressure 1,100 psia

Flow efficiency 0.70 (from pressure transient analysis)

Assume Vogel's inflow performance relationship (IPR) applies. If the flow efficiency is improved to 1.0, the maximum production rate or absolute open flow (STB/day) is most nearly:

a) 420

b) 525

c) 600

d) 860

190. A two-phase reservoir at steady-state is producing at a rate of 640 stb/d with reservoir pressure of 4500 psi. The well is stimulated, resulting in a skin being reduced from 20 to 2. All other properties are shown below:

K = 25 md; h = 120', Bo = 1.3 bbl/stb; μ = 1.7 cp; re = 1490'; rw = 0.406

What is the new production rate for this reservoir for the same bottom-hole flowing pressure?

a) 1250

b) 1450

c) 1750

d) 1550

191. A gas reservoir with re = 1490'; k = 50 md; h = 100'; T=180 °F; and constant reservoir pressure of 2,000 psi has a skin of 26. The gas has an average μ=0.025 cp, and average compresiblity factor of 0.95. Given wellbore radius of 0.328' and bottom-hole flowing pressure of 1,000 psi. What is the production rate of this reservoir?

a) 10,000 Mscf/d

b) 15,000 Mscf/d

c) 20,000 Mscf/d

d) 25,000 Mscf/d

192. A pumping unit is to be installed on a 5,000 ft (1,524 m) well. The foreman wishes to use 1-in (25.4 mm), 7/8-in. (22.2 mm) and .-in. (19.1 mm) sucker rods

to make up the string so he can use his existing inventory and minimize the purchase of new rods. Unfortunately, he only has a limited number of 1-in. (25.4 mm) sucker rods and he will have to buy some of these. What total length, feet (meters), of 1-in. (25.4 mm) sucker rods must be installed in this well if a 2-in. (50.8 mm) pump plunger is needed to achieve the production rate required on this well?

a) 1,435 (437.4)
b) 1,635 (498.3)
c) 1,930 (588.3)
d) 2,260 (688.8)

193. For a well, a pump with a 1-inch plunger is set on 4560 ft of inch rods. The fluid level is known to be low, and the tubing is anchored. With pumping at 20.5 spm and a 64-inch stroke length, production is 355 bbl per day of fluid of specific gravity 0.87. The plunger stroke length is 57.7 inch. The volumetric efficiency (E) for the pump is:

a) 0.60
b) 0.72
c) 0.82
d) 0.50

Statement for Linked Answer Questions 194-197

A 6×18-inch (liner size×stroke length) duplex double acting pump has a volumetric efficiency of 87%. The rod size is the standard 2 inch. The pump rate is 48 spm. The 10,000-ft, $4^1/2$-inch drillstring has a capacity of 0.01422 bbl/ft. The $4^1/2 \times 8^1/2$ inch annulus (10000ft) has a capacity of 0.05 bbl/ft. Neglect the drill collars.

194. The volumetric output in bbl/strokes is:, and the strokes and minutes to displace the annulus and the drill string is:

a) 0.172 bbl/strokes
b) 0.212 bbl/strokes
c) 0.350 bbl/strokes
d) 0.310 bbl/strokes

195. The volumetric output per minute is:

a) 9.25 bbl/minute
b) 8.25 bbl/minute
c) 7.5 bbl/minute
d) 6.0 bbl/minute

196. The number of strokes for the annulus displacement is:

a) 2906 strokes
b) 2510 strokes
c) 1800 strokes
d) 3300 strokes

197. The number of strokes to displace the fluid in drill string is:

a) 750 strokes
b) 655 strokes
c) 825 strokes
d) 910 strokes

Statement for Linked Answer Questions 198-200

A production well produces at rate of 108 STB/D with a bottom hole pressure of 1980 psia. The bubble point pressure of the recombined reservoir fluid to be 1825 psia at a temperature of 195°F. An initial reservoir pressure of 3620 psia was recorded during 48 hour build up following the flow test.

198. The unsaturated productivity index for the above well is:

a) 0.066 stb/psi-day

b) 0.075 stb/psi-day

c) 0.041 stb/psi-day

d) 0.011 stb/psi-day

199. The oil rate if the bottom hole pressure is held at bubble point is:

a) 150 stb/day

b) 118 stb/day

c) 102 stb/day

d) 160 stb/day

200. The maximum permissible oil rate is:

a) 190 stb/day

b) 179 stb/day

c) 150 stb/day

d) 120 stb/day

201. The following information are given on a rod pumped well:

Pumping unit: API 228-173-100 (operating in the long hole)

Pumping speed: 7 SPM

Tubing: 2 7/8 inches (2.441 inches I.D.)

Pump: 2.25-inch tubing pump

Pump depth: 1,000 ft

Rod string: 1-inch steel rods

Production rate:

Oil 351 bbl/d

Water 20 bbl/d

Assuming no significant rod stretch, the volumetric efficiency of this lift system is most nearly:

a) 90%

b) 85%

c) 76%

d) 52%

202. The following information are given for a continuous gas lift well:

Well depth = 8000 ft

Tubing size = $2^3/8$ inch OD

Flowing wellhead pressure = 100 psig

Static bottomhole pressure = 2650 psi

Solution GOR = 200 scf/bbl

Specific gravity of injection gas = 0.07

Surface API gravity of crude oil = 40°

Surface operating pressure = 900 psi

Bottom hole temperature = 200°F

Flowing surface temperature = 120°F

Total GLR = 600 scf/bbl

Point of gas injection = 4625 ft

The amount of gas needed to produce 1000 bbl/d of oil will be:

a) 490.123 scf/d

b) 510.216 scf/d

c) 459.131 scf/d

d) 412.325 scf/d

Statement for Linked Answer Questions 203-207

In an Indian oil field, all gas lift wells have casing pressure operated pressure charged valves. The operating valve is located at 6000 ft.

The pressure in the dome is 700psi and tubing pressure is 500psi at valve depth. The bellow area is 1.0 sq-inch and port size is 0.1 sq-inch. The dome pressure at 60°F is 590 psi.

203. The casing pressure at valve depth required to open the valve is:

a) 716 psi
b) 816 psi
c) 616 psi
d) 636 psi

204. Tubing effect factor is:

a) 0.222
b) 0.111
c) 0.333
d) 0.444

205. Tubing Effect will be:

a) 61.72 psi
b) 71.62 psi
c) 56.72 psi
d) 81.56 psi

206. Spread of valve is:

a) 22.22 psi
b) 31.2 psi
c) 42.4 psi
d) 44.4 psi

207. Test rack opening pressure of valve is:

a) 655 psi
b) 525 psi
c) 725 psi
d) None of the above

Statement for Linked Answer Questions 208-212

The following information are given for a intermittent gas lift well:

Available pressure = 600 psig

Productivity index = 0.3 bbl/day/psi

Well full of water, gradient = 0.5 psi/ft

Spring tension = 100 psi

Gas gravity = 0.6

Full production = 100 bbl/day

Tubing size = 2 inch

Intermittent unloading gradient = 0.04 psi/ft

A_v/A_b = 0.11

Valve opening pressures = 550, 525, 500, 475, and 450 psig

208. The depth of first gas lift valve is:

a) 1400 ft
b) 1200 ft
c) 1600 ft
d) 1800 ft

209. The depth of second gas lift valve is:

a) 2500 ft
b) 2600 ft
c) 2154 ft
d) 2352 ft

210. The depth of third gas lift valve is:

a) 2982 ft
b) 3250 ft
c) 3600 ft
d) 2500 ft

211. The depth of fourth gas lift valve is:

a) 3425 ft

b) 3326 ft

c) 3856 ft

d) 3694 ft

212. The depth of fifth gas lift valve is:

a) 4298 ft

b) 4500 ft

c) 4300 ft

d) 4100 ft

Statement for Linked Answer Questions 213-218

The following information is given for a continuous gas lift well:

Well depth = 8000 ft

Tubing size = $2^7/8$ inch OD

Field fluid gradient = 0.4 psi/ft

Design gas gradient = 0.14 psi/ft

Kick off pressure = 1000 psi

Surface operating pressure = 900 psi

Wellhead pressure = 120 psi

Surface opening pressure of gas lift valve 2, 3, 4, 5, and 6 are 1000, 975, 950, 925 and 900 psi respectively.

213. The setting depth of first gas lift valve is:

a) 2220 ft

b) 2300 ft

c) 5500 ft

d) 2842 ft

214. The setting depth of second gas lift valve is:

a) 3250 ft

b) 3630 ft

c) 3100 ft

d) 3000 ft

215. The setting depth of third gas lift valve is:

a) 4100 ft

b) 4250 ft

c) 4497 ft

d) 4685 ft

216. The setting depth of fourth gas lift valve is:

a) 4212 ft

b) 4498 ft

c) 4785 ft

d) 4623 ft

217. The setting depth of fifth gas lift valve is:

a) 5261 ft

b) 5523 ft

c) 5689 ft

d) 5759 ft

218. The setting depth of fifth gas lift valve is:

a) 5500 ft

b) 5700 ft

c) 5370 ft

d) 5589 ft

Statement for Linked Answer Questions 219-221

For an intermittent gas lift oil well, following information is given:

Casing size = $5^1/2$ inch

Tubing size = 2^12 inch

Gas gravity = 0.70

Flowing THP = 100 psi

Well fluid gradient = 0.40 psi/ft

Injection gas pressure = 700 psi

Valve depth = 4500 ft

Tubing pressure at valve at the time of start of lift cycle = 600 psi

Tubing volume factor = 0.00387 bbl/ft

Slippage loss factor = 0.07/1000 ft of lift

219. The intermittent fluid cycle recovery (vol/cycle) for this well is:
   a) 3.313
   b) 4.414
   c) 5.161
   d) 6.326

220. The number of cycles for this well is:
   a) 210
   b) 150
   c) 106
   d) None of the above

221. Daily production rate from this well is:
   a) 331.3
   b) 431.4
   c) 425.
   d) None of the above

Statement for Linked Answer Questions 222-224

A well produces 100 STB/D oil at a measured flowing bottomhole pressure (BHP) of 1500 psi. A recent pressure survey showed that average reservoir pressure is 2000 psi. Logs indicate a net thickness of 10 ft. The well drains an area with drainage radius, r_e of 1000 ft; the borehole radius is 0.25 ft. Fluid samples indicate that, at current reservoir pressure, oil viscosity is 0.5 cp and formation volume factor is 1.5 RB/STB.

222. The productivity index for the this well is:
   a) 0.2 stb/psi-d
   b) 0.8 stb/psi-d
   c) 1.2 stb/psi-d
   d) 0.5 stb/psi-d

223. Average permeability from the above mentioned data is:
   a) 26 md
   b) 75 md
   c) 16 md
   d) 56 md

224. Skin factor for this well is:
   a) +16
   b) –16
   c) +35
   d) –26

Statement for Linked Answer Questions 225-232

The following information is given for sucker rod pumping well:

Estimate peak and minimum polished rod loads, counterbalance required and peak torque for both Mark II and conventional units for the following conditions:

Pumping depth = 5900 ft

Desired fluid production = 150 bpd

Volumetric efficiency = 80%

Stroke length = 64 inch

Pumping speed = 16.5 spm

Pump diameter = 1 1/4inch

Rod number = API No. 76 (1.814 lb/ft)

Fluid specific gravity = 1.0

225. The peak polished rod load for Mark II unit is:
    a) 14061 lb
    b) 1300 lb
    c) 1100 lb
    d) 1600 lb

226. The minimum polished rod load for Mark II unit is:
    a) 4500 lb
    b) 5643 lb
    c) 3200 lb
    d) 6000 lb

227. The required counter balance for Mark II unit is:
    a) 9852 lb
    b) 9600 lb
    c) 9400 lb
    d) 8500 lb

228. The peak torque for Mark II unit is:
    a) 107,000 in-lb
    b) 102,000 in-lb
    c) 110,000 in-lb
    d) 120,000 in-lb

229. The peak polished rod load for conventional unit is:
    a) 15119 lb
    b) 16000 lb
    c) 18000 lb
    d) 12000 lb

230. The minimum polished rod load for conventional unit is:
    a) 6100 lb
    b) 6700 lb
    c) 5200 lb
    d) 4800 lb

231. The required counter balance for conventional unit is:
    a) 12000 lb
    b) 11000 lb
    c) 10910 lb
    d) 15000 lb

232. The peak torque for conventional unit is:
    a) 173900 in-lb
    b) 153900 in-lb
    c) 143900 in-lb
    d) 193900 in-lb

233. In a gas lift well, the gas is injected at the rate of 1.2 x 10^5 SCF/d and surface injection pressure is 1330 psi. The compressor inlet pressure is 100 psi, and then horse power requirement will be_______
    a) 10 hp
    b) 18.4 hp
    c) 20 hp
    d) 50 hp

234. The pump speed needed to produce 250 STB/d oil at the surface with a rod pump having a 2 in. Diameter plunger, a 50 in. Effective plunger stroke length and a volumetric efficiency of 0.8 will be ------. The oil formation volume factor is 1.2

a) 16 spm

b) 20

c) 30

d) 12.4

235. For the flow of 2000 bbl/d of oil and 1MMSCF/d of gas at a flowing tubing pressure of 800 psi , the choke diameter in 64th of an inch using Gilbert correlation would be ---

a) 30

b) 32

c) 33

d) 35

236. A well is producing from a saturated reservoir with an average reservoir pressure of 2500 psig. Stabilized production test data indicated that the stabilized rate and wellbore pressure are 350 STB/day and 2000 psig, respectively. The Oil flow rate at P_{wf} =1850 psig will be:

a) 1067.1 STB/day

b) 1167.1 STB/day

c) 1267.2 STB/day

d) 1310.5 STB/day

237. Which of the following are the causes of sand production?

a) Totally unconsolidated formations

b) High production rate

c) Reservoir Depletion

d) All of the above

238. Which of the following problem is associated with sand production in the oilfields?

a) Wear of downhole and surface equipments and casing collapse

b) Borehole instability

c) Workover and separation costs

d) All of the above

239. Which of the following method is used to control sand production problem?

a) Mechanical sand control

b) Downhole sand consolidation

c) Downhole emulsification

d) All of the above

240. Which of the following statement is true about sand production in the oilfields?

a) Sand production problems are encountered during clean up after perforating or acidizing

b) Sand is produced from unconsolidated or poorly cemented sandstone reservoirs

c) Sand production occurs when the reservoir fluids are excessively produced

d) All of the above

241. The sand production during water breakthrough in secondary oil recovery operations falls under the category of:

a) Transient sand production

b) Continuous sand production

c) Catastrophic sand production

d) None of the above

242. The sand production encountered during excessive production of reservoir fluids falls under the category of:
   a) Transient sand production
   b) Continuous sand production
   c) Catastrophic sand production
   d) None of the above

243. Sand production from unconsolidated sandstone reservoirs having no sand control equipment falls under the category of:
   a) Transient sand production
   b) Continuous sand production
   c) Catastrophic sand production
   d) None of the above

244. Which of the following is true about sieve analysis
   a) It involves sorting of sand grains of similar sizes using a series of sieves?
   b) Prior to conducting sieve analysis, sample should be cleaned and dried
   c) Both (a) and (b)
   d) None of the above

245. Uniformity coefficient is determined by
   a) Comparing 40 percentile size with 90 percentile size
   b) Comparing 40 percentile size with 80 percentile size
   c) Comparing 40 percentile size with 70 percentile size
   d) None of the above

246. Sand is uniform and described by the (D_{10}) size when
   a) Uniformity coefficient is less than 5
   b) Uniformity coefficient is greater than 5
   c) Uniformity coefficient is greater than 10
   d) None of the above

247. Sand is non uniform and described by the (D_{40}) size when
   a) Uniformity coefficient is less than 5
   b) Uniformity coefficient is greater than 5
   c) Uniformity coefficient is greater than 10
   d) None of the above

248. Sand is very non uniform and described by the (D_{70}) size when
   a) Uniformity coefficient is less than 5
   b) Uniformity coefficient is greater than 5
   c) Uniformity coefficient is greater than 10
   d) None of the above

249. As per Schwartz method, which of the following criteria of gravel selection is considered for non- uniform sand and with flow velocity greater than 0.05 ft/s?
   a) D_{10} gravel = 6 times D_{10} sand
   b) D_{40} gravel = 6 times D_{40} sand
   c) D_{50} gravel = 6 times D_{50} sand
   d) None of the above

250. As per Schwartz method, which of the following criteria of gravel selection is considered for uniform sand and with flow velocity less than 0.05 ft/s?

a) D_{10} gravel = 6 times D_{10} sand

b) D_{10} gravel = 6 times D_{10} sand

c) D_{10} gravel = 6 times D_{10} sand

d) None of the above

251. Which of the following statement is true as per Saucier's Design Criteria

a) The gravel should be selected on the basis of 40 percentile grain size of smallest productive formation sand of the core within the interval that is to be gravel packed.

b) The selected 50 percentile size of the gravel should be less than six times the 50 percentile size of the of the smallest grained productive formation sand size

c) The selected 40 percentile size of the gravel should be less than six times the 40 percentile size of the of the smallest grained productive formation sand size

d) None of the above

252. Which of the following is an example of major workover job?

a) Reperforation

b) Installation of artificial lift system

c) Replacement of leaking tubing

d) None of the above

253. Which of the following is an example of routine workover job?

a) Stimulation

b) Installation of artificial lift system

c) Plug and abandon

d) None of the above

254. ________is the process of running or pulling tubing, drill pipe or other tubular in the presence of sufficient surface pressure present to cause the tubular to be foced out of the hole.

a) Snubbing

b) Perforating

c) Drilling

d) None of the above

255. Which of the following is the main cause of excessive water production in an oil well?

a) Channelling

b) Coning

c) Both (a) and (b)

d) None of the above

256. ________is the attraction of water from an adjacent connecting zone towards the wellbore due to reduced pressure caused by production

a) Water coning

b) Water fingering

c) Water channelling

d) None of the above

257. Channelling occurs when

a) High permeability channels exist between a water source and a production well

b) Low permeability channels exist between a water source and a production well

c) Both (a) and (b)

d) None of the above

258. Fingering occurs when

a) The mobility of the displacing fluid is higher than that of the displaced fluid

b) The mobility of the displacing fluid is lower than that of the displaced fluid

c) Both (a) and (b)

d) None of the above

259. The technology intends to reduce water production is called

a) Water shut-off

b) Conformance control

c) Both (a) and (b)

d) None of the above

260. The reservoir pressure is 2700 psi and the vertical depth of the exposed perforations is 5000'. The required fluid density to kill the well will be:

a) 10.4 ppg

b) 9.4 ppg

c) 11.4 ppg

d) 12.4 ppg

261. The point where the tubing starts to buckle is known as:

a) Neutral Point

b) Critical Point

c) Middle Point

d) None of the above

262. If holes in the casing are found by inspection tools what can be done to repair?

a) Tubing patch

b) Liner

c) Casing patch

d) Drilling jar

Answers

Q1.	a	Q36.	b	Q71.	a	Q106.	c	Q141.	c
Q2.	d	Q37.	a	Q72.	a	Q107.	b	Q142.	d
Q3.	a	Q38.	b	Q73.	c	Q108.	c	Q143.	b
Q4.	a	Q39.	a	Q74.	d	Q109.	b	Q144.	a
Q5.	c	Q40.	b	Q75.	a	Q110.	a	Q145.	c
Q6.	d	Q41.	c	Q76.	a	Q111.	c	Q146.	a
Q7.	d	Q42.	a	Q77.	a	Q112.	d	Q147.	a
Q8.	d	Q43.	a	Q78.	c	Q113.	c	Q148.	a
Q9.	c	Q44.	d	Q79.	d	Q114.	d	Q149.	a
Q10.	c	Q45.	b	Q80.	d	Q115.	d	Q150.	b
Q11.	c	Q46.	a	Q81.	d	Q116.	b	Q151.	a
Q12.	d	Q47.	c	Q82.	d	Q117.	a	Q152.	b
Q13.	a	Q48.	b	Q83.	d	Q118.	a	Q153.	c
Q14.	a	Q49.	c	Q84.	d	Q119.	b	Q154.	c
Q15.	b	Q50.	a	Q85.	d	Q120.	d	Q155.	d
Q16.	c	Q51.	a	Q86.	c	Q121.	c	Q156.	a
Q17.	d	Q52.	a	Q87.	c	Q122.	c	Q157.	b
Q18.	c	Q53.	a	Q88.	a	Q123.	c	Q158.	d
Q19.	c	Q54.	b	Q89.	d	Q124.	d	Q159.	a
Q20.	c	Q55.	a	Q90.	c	Q125.	b	Q160.	b
Q21.	d	Q56.	c	Q91.	c	Q126.	a	Q161.	b
Q22.	a	Q57.	b	Q92.	d	Q127.	b	Q162.	b
Q23.	b	Q58.	a	Q93.	d	Q128.	b	Q163.	a
Q24.	d	Q59.	a	Q94.	b	Q129.	c	Q164.	b
Q25.	a	Q60.	b	Q95.	a	Q130.	c	Q165.	c
Q26.	c	Q61.	a	Q96.	b	Q131.	b	Q166.	b
Q27.	a	Q62.	b	Q97.	a	Q132.	b	Q167.	a
Q28.	a	Q63.	b	Q98.	b	Q133.	a	Q168.	a
Q29.	b	Q64.	a	Q99.	d	Q134.	b	Q169.	a
Q30.	b	Q65.	a	Q100.	b	Q135.	a	Q170.	a
Q31.	b	Q66.	d	Q101.	c	Q136.	a	Q171.	d
Q32.	b	Q67.	a	Q102.	d	Q137.	b	Q172.	a
Q33.	a	Q68.	b	Q103.	a	Q138.	b	Q173.	b
Q34.	c	Q69.	c	Q104.	b	Q139.	b	Q174.	a
Q35.	a	Q70.	a	Q105.	a	Q140.	c	Q175.	b

Answers									
Q176.	b	Q194.	a	Q212.	a	Q230.	b	Q248.	c
Q177.	a	Q195.	b	Q213.	a	Q231.	c	Q249.	b
Q178.	d	Q196.	a	Q214.	b	Q232.	b	Q250.	a
Q179.	a	Q197.	c	Q215.	c	Q233.	b	Q251.	b
Q180.	b	Q198.	a	Q216.	b	Q234.	a	Q252.	a
Q181.	b	Q199.	b	Q217.	a	Q235.	c	Q253.	b
Q182.	a	Q200.	b	Q218.	c	Q236.	a	Q254.	a
Q183.	c	Q201.	a	Q219.	a	Q237.	d	Q255.	c
Q184.	b	Q202.	c	Q220.	c	Q238.	d	Q256.	a
Q185.	a	Q203.	a	Q221.	a	Q239.	d	Q257.	a
Q186.	b	Q204.	b	Q222.	a	Q240.	d	Q258.	a
Q187.	a	Q205.	a	Q223.	c	Q241.	a	Q259.	c
Q188.	b	Q206.	a	Q224.	a	Q242.	c	Q260.	a
Q189.	d	Q207.	a	Q225.	a	Q243.	b	Q261.	a
Q190.	c	Q208.	b	Q226.	b	Q244.	c	Q262.	c
Q191.	c	Q209.	c	Q227.	a	Q245.	a		
Q192.	a	Q210.	a	Q228.	a	Q246.	a		
Q193.	c	Q211.	d	Q229.	a	Q247.	b		

CHAPTER **5**

Petroleum Production Operations-II

1. Which of the following theory is used to understand the separation of oil and gas in a separator?
   a) Daltons theory
   b) Droplet settling theory
   c) Bohrs Theory
   d) None of the above

2. What are the main principles of physical separation of gas and liquids in a separator?
   a) Momentum Change
   b) Gravity Settling
   c) Coalescence
   d) All of the above

3. For low gas oil ratio, the most suitable separator is
   a) Vertical separator
   b) Horizontal Separator
   c) Spherical separator
   d) None of the above

4. Horizontal separators are used for
   a) Low gas oil ratio
   b) High gas oil ratio
   c) Both of these
   d) None of the above

5. The main constituent of natural gas is
   a) Methane
   b) Carbon dioxide
   c) Carbon monoxide
   d) Nitrogen dioxide

6. The function of mist extractor is to remove
   a) Very fine liquid droplets from gas
   b) Very fine water droplets from oil
   c) Very fine oil droplets from water
   d) None of the above

7. ________ are used to cause the initial bulk separation of liquid and gas in a separator
   a) Inlet divertor
   b) Wave breaker
   c) Defoaming plates
   d) None of the above

8. Wave breakers are generally used in
   a) Horizontal separators
   b) Vertical separators
   c) Both (a) and (b)
   d) None of the above

9. Which of the following is the main advantage of weir plate used in horizontal separator?
   a) Interfacial (oil-water level) control is easier to operate
   b) Relative retention times can be varied easily
   c) The cleanest oil is taken off the top
   d) All of the above

10. Which of the following is the main disadvantage of weir plate used in horizontal separator?
   a) Use of baffles make removal of sand and mud difficult
   b) A considerable portion of vessel volume is vested
   c) Both (a) and (b)
   d) None of the above

11. Which of the following is the main advantage of open separator?
   a) Relative retention volumes can be varied easily
   b) No traps to accommodate sand, mud etc
   c) Both (a) and (b)
   d) None of the above

12. Which of the following is the main advantage of bucket and weir type separators?
   a) Only float level control is used and controllers sense the large density difference between liquid and gas
   b) If level control or valves fail, only that liquid and gas are discharged
   c) Both (a) and (b)
   d) None of the above

13. Which of the following is true about the disadvantage of bucket and weir type separator
   a) Difficult to remove sand and mud
   b) More vessel space is vested
   c) Both (a) and (b)
   d) None of the above

14. For the same design capacity, ____ perform better for higher GOR flow stream.
   a) Horizontal Separators
   b) Vertical Separators
   c) All of the above
   d) None of the above

15. An oil emulsion is __________?
   a) A heavy viscous liquid
   b) A heavy viscous liquid containing a large amount of entrained sediment
   c) An oil & water mixture that does not readily separate
   d) A layer of free water located above a heavy viscous petroleum product

16. What are the main criteria that are necessary for formation of crude oil emulsion?
   a) Two immiscible liquids must be brought in contact
   b) Surface active component must present as the emulsifying agent
   c) Sufficient mixing and agitation effect must be provided in order to disperse one liquid into another as droplets
   d) All of the above

17. Which of the following influence the rate of breakdown process in emulsions?
    a) Flocculation
    b) Sedimentation
    c) Coalescence
    d) All of the above

18. Which of the following method is used for the treatment of emulsion in crude oil?
    a) Application of heat
    b) Application of chemicals
    c) Application of electricity
    d) All of the above

19. Which of the following operating problem may be caused by salt present in the crude oil?
    a) Salt cakes out inside the equipments and causes poor flow and plugging, reduce heat transfer rates in exchangers and causes high heater tube wall temperature and plugged fractionators tray
    b) It poisons the catalyst used in resid cracking and hydrotreating
    c) Both (a) and (b)
    d) None of the above

20. The low temperature separators are used for
    a) The removal of light condensable hydrocarbons from high pressure gas stream
    b) The removal of heavy condensable hydrocarbons from high pressure streams
    c) Both (a) and (b)
    d) None of the above

21. The temperature reduction in low temperature separator is obtained by
    a) Joule Thomson Effect
    b) Peltier Effect
    c) Gravity Effect
    d) None of the above

22. In order to maximize the hydrocarbon liquid recovery, which of the following separation mechanism is used?
    a) Flash separation
    b) Differential separation
    c) Both (a) and (b)
    d) None of the above

23. Field processing of natural gas consists of :
    a) Separation of gas from free liquids and processing of gas to remove condensate and recoverable hydrocarbon vapours
    b) Processing of gas to remove condensable water vapour, which may cause hydrate formation
    c) Processing of gas to remove undesirable components like hydrogen sulphide and carbon dioxide
    d) All of the above

24. In ________, light gases are separated due to reduction in pressure?
    a) Flash separation
    b) Differential separation
    c) Both (a) and (b)
    d) None of the above

25. In ______, gases liberated from oil are kept in intimate contact with liquid phase and thermodynamic equilibrium is established between the two phases and separation take place at the required pressure.
    a) Flash separation
    b) Differential separation
    c) Both (a) and (b)
    d) None of the above

26. How many stages are required for flash separation?
    a) 2-4 stages
    b) 20-40 stages
    c) 40-50 stages
    d) Too many stages

27. How many stages are required for differential separation?
    a) 2-4 stages
    b) 20-40 stages
    c) 40-50 stages
    d) Too many stages

28. The orifice plate in a gas meter should be installed with the sharp edge:
    a) upstream
    b) downstream
    c) parallel to the flow stream
    d) it does not matter

29. Which of the following equation is used for the determination of the diameter of pipeline?
    a) $D = 0.422\ Q^{0.4}\ (fSL/\Delta P)^{0.2}$ where D is the diameter of the pipeline in inch, ΔP is the pressure drop in psi, f is friction factor, S is specific gravity, L is length of pipeline in ft, Q flow rate in gpm
    b) $D = 0.422\ Q^{0.2}\ (fSL/\Delta P)^{0.4}$ where D is the diameter of the pipeline in inch, ΔP is the pressure drop in psi, f is friction factor, S is specific gravity, L is length of pipeline in ft, Q flow rate in gpm
    c) $D = 0.422\ Q\ (fSL/\Delta P)^{0.2}$ where D is the diameter of the pipeline in inch, ΔP is the pressure drop in psi, f is friction factor, S is specific gravity, L is length of pipeline in ft, Q flow rate in gpm
    d) $D = 0.422\ Q^{0.2}\ (fSL/\Delta P)^{0.5}$ where D is the diameter of the pipeline in inch, ΔP is the pressure drop in psi, f is friction factor, S is specific gravity, L is length of pipeline in ft, Q flow rate in gpm

30. Which of the following equation is used to calculate erosional velocity?
    a) $V_e = 100/\rho^2$ where V_e is the erosional velocity in ft/s, ρ is specific gravity in fps system
    b) $V_e = 100/\rho^3$ where V_e is the erosional velocity in ft/s, ρ is specific gravity in fps system
    c) $V_e = 100/\rho^{0.5}$ where V_e is the erosional velocity in ft/s, ρ is specific gravity in fps system
    d) $V_e = 100/\rho^{0.25}$ where V_e is the erosional velocity in ft/s, ρ is specific gravity in fps system

31. The recommended velocity of water in pipelines is
    a) 3-10 ft/s
    b) 50-75 ft/s
    c) 75-100 ft/s
    d) None of the above

32. Which of the following method is used for heavy crude oil and bitumen transportation via pipeline?
    a) Viscosity reduction method
    b) In Situ upgrading method
    c) Both (a) and (b)
    d) All of the above

33. Which of the following friction reduction technique is used for heavy oil transportation?
    a) Use of drag reducing additives
    b) Core annular flow
    c) Both (a) and (b)
    d) None of the above

34. Which of the following viscosity method is used for heavy oil transportation?
    a) Dilution
    b) Blending
    c) Emulsificaion
    d) All of the above

35. Which of the following method is used to reduce viscosity and improve flow behavior of heavy crude oil?
    a) Pour point reduction
    b) Heating
    c) Both (a) and (b)
    d) None of the above

36. Which of the following evaporative emission is associated with large storage tanks?
    a) Breathing loss
    b) Working and standing storage loss
    c) Both (a) and (b)
    d) None of the above

37. Which of the following evaporative emission is associated with Motor vehicle tanks?
    a) Refueling loss
    b) Working and standing storage loss
    c) Breathing loss
    d) Ballasting loss

38. Which of the following evaporative emission is associated with service stations?
    a) Bulk fuel drop losses
    b) Underground storage breathing loss
    c) Both (a) and (b)
    d) None of the above

39. Which of the following evaporative emission is associated with rail tank cars, tank trucks and marine vessels?
    a) Bulk fuel drop losses
    b) Underground storage breathing loss
    c) Transit loss
    d) None of the above

40. __________ are the primary source of evaporative emission is associated with rail tank cars, tank trucks and marine vessels?

a) Loading losses

b) Breathing Losses

c) Both (a) and (b)

d) None of the above

41. Which of the following method is used for cargo carrier loading?

a) Splash loading

b) Submerged Loading

c) Bottom loading

d) All of the above

42. Assume that 2.0 MMscf/d of natural gas is saturated with water vapor at 100°F and 1,000 psig. The amount of water vapor (pounds) that must be removed per day to condition the gas to sales line specification (7 lb/MMcf) is most nearly:

a) 14

b) 58

c) 116

d) 232

43. Which of the following flow meter is used for custody transfer of petroleum and petroleum products?

a) Orifice flow meter

b) Coriolis flow meter

c) Turbine Meter

d) All of the above

44. Which of the following flow meter measures mass flow rate?

a) Coriolis flow meter

b) Turbine flow meter

c) Positive Displacement flow meter

d) None of the above

45. Which of the following flow meter is based on Bernoulli's Theorem?

a) Orifice meter

b) Coriolis flow meter

c) Sonic flow meter

d) None of the above

46. __________ has a semicircular opening and is used for liquids with suspended solids or high concentration of entrained gas

a) Square edged concentric bore orifice plate

b) Eccentric bore orifice plate

c) Segmented orifice plate

d) None of the above

47. __________ has one hole located near the top for liquids and near the bottom of the pipe for gases

a) Concentric bore orifice plate

b) Segmented orifice plate

c) Both of these

d) None of the above

48. Which of the following type of tape is used in orifice flowmeter?

a) D and D/3 Tapes

b) D and D/2 tapes

c) D and D/4 tapes

d) None of the above

49. Which of the following system is not used in the pressure taping of orifice flowmeter?

a) D and D/4 taps

b) Corner tapes

c) Flange Tapes

d) Both (b) and (c)

50. Which of the following is the practical orifice meter flow equation?

a) $q_m = N_1 C_d E_v \gamma d^{0.2} \sqrt{\rho_{t,p} \Delta p}$

where C_d is orifice plate co-efficient of discharge, d is orifice plate diameter, Δp orifice differential pressure, E_v is velocity of approach, q_m is the mass flow rate, $\rho_{t,p}$ is density of fluid at flowing temperature and flowing pressure, γ is the expansion factor

b) $q_m = N_1 C_d E_v \gamma d^2 \sqrt{\rho_{t,p} \Delta p}$

where C_d is orifice plate co-efficient of discharge, d is orifice plate diameter, Δp orifice differential pressure, E_v is velocity of approach, q_m is the mass flow rate, $\rho_{t,p}$ is density of fluid at flowing temperature and flowing pressure, γ is the expansion factor

c) $q_m = N_1 C_d E_v \gamma d^{0.25} \sqrt{\rho_{t,p} \Delta p}$

where C_d is orifice plate co-efficient of discharge, d is orifice plate diameter, Δp orifice differential pressure, E_v is velocity of approach, q_m is the mass flow rate, $\rho_{t,p}$ is density of fluid at flowing temperature and flowing pressure, γ is the expansion factor

d) $q_m = N_1 C_d E_v \gamma d^3 \sqrt{\rho_{t,p} \Delta p}$

where C_d is orifice plate co-efficient of discharge, d is orifice plate diameter, Δp orifice differential pressure, E_v is velocity of approach, q_m is the mass flow rate, $\rho_{t,p}$ is density of fluid at flowing temperature and flowing pressure, γ is the expansion factor

51. The main method of natural gas transportation is

a) Pipeline

b) Tankers

c) Truck

d) None of the above

52. The factors affecting the flow condition in two phase pipe-lines?

a) Gas-liquid ratio

b) Velocity of gas and liquid

c) Fluid characteristics

d) All of the above

53. As the density of a material increases, the API Gravity?

a) becomes higher

b) becomes lower

c) is not changed

d) none of the above

54. In what units is a metric tape graduated?

a) Millimeters

b) Milliliters

c) Hundredths of a foot

d) Percentages

55. API gravity of a liquid having sp.gr of 0.876 is
    a) 30°API
    b) 36°API
    c) 33°API
    d) 42°API

56. The API gravity of water at 60°F is?
    a) 6
    b) 10
    c) 15
    d) 1.0

57. Crude oil that has a high API gravity, usually more than 40 degrees is
    a) Light crude oil
    b) Heavy crude oil
    c) Extra heavy crude oil\
    d) All of the above

58. BS&W stands for
    a) Basic sediment and water
    b) Basic sand and water
    c) Basic scale and water
    d) None of the above

59. _______ are generally used to store products that don't readily vaporize at ambient temperature or stored temperature conditions
    a) Fixed roof tanks
    b) Floating roof tanks
    c) Bullet tanks
    d) None of the above

60. ________are generally used to store the products that readily vaporize at ambient temperature or stored temperature
    a) Open tanks
    b) Fixed roof tanks
    c) Floating roof tanks
    d) None of the above

61. Which of the following floating roof tank consists of a fixed roof at the top of the shell?
    a) Internal floating roof tank
    b) Domed external floating roof tank
    c) Both of these
    d) None of the above

62. The floating roof tank is used for
    a) Gasoline
    b) Lubricating oil
    c) Kerosene
    d) All of the above

63. Which of the following tank can be used for liquid petroleum gas?
    a) Fixed roof tank
    b) Floating roof tanks
    c) Bullet tanks
    d) None of the above

64. ______ are used to store the products having very high vapour pressure
    a) Spherical Tanks
    b) Fixed roof tanks
    c) Floating roof tanks
    d) None of the above

65. Which of the following tank can be used for propane?
   a) Bullet Tanks
   b) Fixed roof tanks
   c) Floating roof tanks
   d) None of the above

66. Class I hydrocarbons are those liquid crude oil derivatives which have _______
   a) Flash point below 21°C
   b) Flash point from 21°C to 55°C
   c) Flash point above 55°C
   d) None of the above

67. Class II hydrocarbons are those liquid crude oil derivatives which have _______
   a) Flash point below 21°C
   b) Flash point from 21°C to 55°C
   c) Flash point above 55°C
   d) None of the above

68. Class III hydrocarbons are those liquid crude oil derivatives which have _______
   a) Flash point below 21°C
   b) Flash point from 21°C to 55°C
   c) Flash point above 55°C and upto 100°C
   d) None of the above

69. Which of the following storage tank is recommended for Class I hydrocarbons?
   a) Floating Roof Tanks
   b) Fixed Roof tanks
   c) Both (a) and (b)
   d) None of the above

70. Which of the following storage tank is recommended for Class II hydrocarbons?
   a) Floating Roof Tanks
   b) Fixed Roof tanks
   c) Both (a) and (b)
   d) None of the above

71. Which of the following storage tank is recommended for Class III hydrocarbons?
   a) Floating Roof Tanks
   b) Fixed Roof tanks
   c) Both (a) and (b)
   d) None of the above

72. LNG storage tanks have
   a) single container
   b) Double containers
   c) Triple Containers
   d) None of the above

73. The LNG storage tanks have ability to store LNG at
   a) the very low temperature of -162°C
   b) the temperature of 30°C
   c) the temperature of -30°C
   d) None of the above

74. Which of the following is true about the spacing of floating roof tanks?
   a) The spacing between adjacent floating roof tanks should be 10 metres for the tanks having diameter less than 45 metre

b) The spacing between adjacent floating roof tanks should be 15 metres for the tanks having diameter less than 45 metre for the tanks having diameter greater than 45 metres

c) Both (a) and (b)

d) None of the above

75. Which of the following is true about the spacing of fixed roof tanks?

a) The spacing between adjacent fixed roof tanks should be equal to the diameter of smaller tank

b) The spacing between adjacent fixed roof tanks should be equal to the half of the larger tank

c) Both (a) and (b)

d) None of the above

76. Which of the following is true about the spacing between a floating roof tank and a fixed roof tank?

a) The spacing between a floating roof tank and a fixed roof tank should be equal to the diameter of smaller tank

b) The spacing between a floating roof tank and a fixed roof tank should be equal to the half of the larger tank

c) Both (a) and (b)

d) None of the above

77. The gas blanketing system of a fixed roof tank prevents

a) The evaporation of hydrocarbon products

b) Introduction of air

c) Both (a) and (b)

d) None of the above

78. Where does the blanket gas get vented to when pressure rises in the tank?

a) Vapour recovery system

b) Atmosphere

c) Both (a) and (b)

d) None of the above

79. What safety device is used in the storage tanks to regulate fast pressure changes?

a) Breather valve

b) Check valve

c) Stuffing Box

d) None of the above

80. What is the purpose of smothering steam in a tank

a) To heat the liquid hydrocarbons

b) To combat internal tank fire by displacing air

c) To heat the tank shell

d) None of the above

81. How internal fire of the storage tanks can be prevented?

a) By foam injectors

b) By smothering steam

c) Both (a) and (b)

d) None of the above

82. What is the option tanks have to combat the buildup of electric charge:

a) Rectifier

b) Ground

c) Both (a) and (b)

d) None of the above

83. A running sample is_____

a) Obtained by lowering an unstoppered beaker or bottle from the top of the oil to the level of the outlet (Suction) and returning it to the top of the oil at a uniform rate so that the beaker or bottle is 70% to 85% full when withdrawn from the oil

b) Obtained by lowering a stoppered beaker or bottle to the level of the outlet, then opening the sampler and raising it at a uniform rate so that it is about 50% full when withdrawn from the oil

c) Obtained by lowering a stoppered beaker to the mid point of the tank contents, then opening the sampler and raising and lowering it at a uniform rate until it is full

d) Obtained by lowering an unstoppered beaker or bottle to the mid point of the tank contents, then raising and lowering it at a uniform rate until it is full

84. An all levels sample is obtained by____

a) Submerging an unstoppered beaker or bottle to a point near the tank draw-off (suction) level and then raising it, all at a uniform rate, so that it is no more than 50% full on emerging from the liquid

b) Blending upper, middle and lower samples from the same tank

c) Submerging a stoppered beaker or bottle to a point as near as possible to the draw-off (suction) level, then opening the container and raising it at a rate such that it is 70% to 85% full as it emerges from the liquid

d) Submerge a stoppered beaker or bottle to the mid point of the product in a tank, then opening the sampler and raising and lowering it at a uniform rate until the sampler is full

85. A lower sample is taken from____

a) The bottom of the tank

b) Just above the water level

c) The mid-point of the bottom third of the liquid

d) Any of the above

86. A top sample is taken from____

a) The surface of the liquid

b) 6" below the surface

c) The mid-point of the upper third of the liquid

d) Anywhere in the top portion of the tank

87. Extraordinary care is required when selecting a container for____
   a) Gasoline samples
   b) Fuel oil samples
   c) Benzene samples
   d) Jet fuel samples

88. Rinsing the container with the liquid to be sampled before drawing the sample is recommended for____
   a) Jet fuel samples
   b) Petrochemical samples
   c) Vapor pressure samples
   d) All of the above

89. What equipment is preferred by the API to lower a sample bottle to the required level?
   a) Ullage tape
   b) Cotton cord or non-sparking chain marked to indicate when the correct level has been reached
   c) Innage tape
   d) Brass sampling cage

90. Free water in a crude oil tank is best sampled using a_____
   a) Zone sampler
   b) Bacon bomb or tube-type sampler
   c) Bottle and sample cage
   d) Weighted bottle and cord

91. A sampling cage and bottle is generally better than a weighted sampling beaker for sampling volatile liquids because_____
   a) The equipment is easier to handle
   b) A beaker sample is probably less representative
   c) Loss of light ends is likely when the sample is transferred from the beaker
   d) Sample bottles are readily available

92. A Bacon-Bomb sampler is used to take______
   a) Tank bottom samples
   b) Running samples
   c) LPG samples
   d) All of the above

93. When would you take a free water sample?
   a) When requested by the customer
   b) They are not necessary
   c) Immediately after every loading, except chemicals
   d) Whenever there is a sufficient quantity to sample

94. Free water samples are important for_____
   a) Checking for contaminated water under gasoline
   b) Determining the likely source of the free water
   c) Determining the influence of free water on the API gravity of a crude oil cargo
   d) None of the above

95. A sample obtained by lowering a stoppered container to the drawoff level of a tank, removing the stopper and withdrawing the container at a steady rate is called a___

a) Spot sample
b) Running sample
c) All-levels sample
d) Composite sample

96. A sample obtained by lowering an unstoppered container to the draw-off level of a tank and withdrawing the container without stopping is called a____
a) Multi-level sample
b) Running sample
c) All-levels sample
d) Composite sample

97. When sampling heavy fuel oils or crude petroleum products in a tank with sludge or sediment deposits on the bottom, manual samples may not be representative because____
a) The material may be non-homogeneous
b) The concentration of entrained water is higher near the bottom
c) The interface between the oil and water is difficult to measure accurately
d) All of the above

98. After obtaining a tank running sample, the inspector notices that the sampling bottle is full on withdrawing it from the tank. The inspector should____
a) Pour some of the sample into a second container
b) Pour some of the sample out of the bottle
c) Empty the sampling bottle completely and obtain a new sample
d) Any of the above

99. A storage tank is manually gauged and found to contain 8 feet of product. How many spot samples should be obtained from this tank?
a) 2 [Upper and lower]
b) 3 [Upper, middle and lower]
c) 1 [Middle]
d) None

100. When you are instructed not allow the sampling cord to fall on the deck, the main concern is__________
a) The chance of contamination
b) Making a mess that could cause a slip hazard
c) Wearing out the cord
d) Both a and b

101. A running sample, when taken correctly, must___
a) Be at least half full
b) Taken with a zone sampler
c) Be taken with a brass beaker
d) Be approximately 70% to 85% full

102. When sampling a shore tank with an observed API gravity of -2°, and the main concern is water. Where would the water most likely be?
a) 6" off the bottom
b) Stratified between the lower and middle levels

c) Floating on top of the cargo

d) No water can be found in negative API oils

103. The reason a running sample brought up full is not allowed because_____

a) There is no way to tell when the bottle filled up

b) There would be too much oil to test the gravity

c) There is a chance of contamination from the bottle cap

d) As the bottle warms up, it could shatter

104. The correct procedure for taking a jet fuel sample is____

a) First take a sample in a clear bottle to examine for color and sediment. Then take the official sample an amber bottle

b) Use an amber bottle

c) Use only a clean zone sampler

d) Sample just below the surface as to keep from getting any free water in it

105. What is the principal disadvantage of using a beaker sampler to sample a marine vessel?

a) They are too heavy to hoist up and down

b) If one tank is contaminated, subsequent samples could also become contaminated

c) There is no cap that fits them

d) They are too expensive and might be lost in the tank

106. What is the principal reason for taking 3" and 6" bottom samples in No. 6 Fuel Oil?

a) To determine the gauge height

b) To obtain a sample for sulfur & viscosity determination

c) To locate any free water not identified by water paste

d) Bacon-bomb samples should never be taken of No. 6 Residual Fuel Oil

107. A bottom or dead-bottom sample is taken at what point?

a) Six inches below the outlet

b) Four inches below the outlet

c) At the lowest point in the tank

d) Immediately above the free water level

108. A sample taken at a specific location in a tank is called_____

a) An official sample

b) A tank-side sample

c) A spot sample

d) A single-tank composite sample

109. At what point in the product is a top sample taken?

a) Middle of the upper third of the liquid

b) Middle of the product

c) Middle of the lower third of the tank

d) Six inches below the surface of the liquid

110. A storage tank was manually gauged and found to have a product innage of 38 feet 6 inches. The tank gauge height is 48 feet 10 inches. At what innage level should the lower spot sample be obtained?

a) 3 feet from the tank bottom

b) At the outlet (suction) level

c) 6 feet 5 inches from the tank bottom

d) 12 feet 10 inches from the tank bottom

111. A storage tank was manually gauged and found to have a product innage of 38 feet 6 inches. The tank gauge height is 48 feet 10 inches. At what innage level should the middle product spot sample be obtained?

a) 12 feet 10 inches from the tank bottom

b) 19 feet 3 inches from the tank bottom

c) 24 feet 5 inches from the tank bottom

d) 6 feet 5 inches below the surface of the product

112. A storage tank was manually gauged and found to have a product innage of 38 feet 6 inches. The tank gauge height is 48 feet 10 inches. At what innage level should the upper product spot sample be obtained?

a) 32 feet 1 inch from the tank bottom

b) 5 feet 5 inches below the surface of the product

c) 40 feet 9 inches from the tank bottom

d) 6 inches below the surface of the product

113. An Upper sample is taken from____

a) The surface of the liquid

b) 6" (0.15m) below the surface

c) The mid-point of the upper third of the liquid

d) Anywhere in the top portion of the tank

114. Upper, middle, and lower samples____

a) can be taken in any order

b) must be taken in the order lower, middle, upper

c) must be taken in the order upper, middle, lower

d) are less reliable than an all-levels sample, no matter how they are taken

115. When upper-middle-lower samples are taken from a tank, the lower sample is taken at a point_

a) 1/3 of the liquid height from the tank bottom

b) At the middle of the tank outlet fitting

c) 1/6 of the liquid height from the tank bottom

d) 6" (0.15m) off the tank bottom

116. When drawing upper-middle-lower samples from any tank, which sample should be taken first?

a) Lower

b) Middle

c) Upper

d) It doesn't matter

117. When upper-middle-lower samples are taken from a tank, the upper sample is taken at a point_

a) 1/3 of the liquid height from the tank bottom

b) At the middle of the tank outlet fitting

c) Middle of the upper third of the tank contents

d) 6" (0.15m) off the tank bottom

118. When drawing upper-middle-lower samples from any tank, which sample should be taken last?

a) Lower

b) Middle

c) Upper

d) It doesn't matter

119. When drawing upper-middle-lower samples from any tank, which sample should be taken second?

a) Lower

b) Middle

c) Upper

d) It doesn't matter

120. At what vertical location in the liquid is an upper sample taken?

a) Middle of the upper third of the liquid

b) One inch below the surface of the liquid

c) Six inches (0.15m) below the surface of the liquid

d) Middle of the lower third of the tank

121. An automatic sampling system consists of___

a) Probe

b) Receiver

c) Controller

d) All of the above

122. Which of the following sample are found in the sample taken by an automatic in-line sampling system?

a) Free water

b) Entrained water

c) Both (a) and (b)

d) None of the above

123. A sample that is to be tested for Reid vapor pressure [RVP] should be taken with___

a) A zone sampler

b) A glass bottle with suitable cage or weight

c) A bomb sampler

d) A Tulsa Thief

124. When drawing a running sample for RVP, the sample bottle should be pulled from the liquid?

a) 50% full

b) 70-85% full

c) 90-95% full

d) 100% full

125. Regardless of the product, sample containers must not be filled in excess of ___

a) 50%

b) 85%

c) 60%

d) 100%

126. When a properly-functioning automatic in-line sampler is used during the discharge of a crude oil vessel, the resulting sample will include?

a) Oil and S&W

b) Oil, S&W and sludge

c) Oil, S&W & Free Water

d) S&W and Free Water only

127. Samples from each vessel tank must be composited____

a) On board as long as every cargo tank contains the same cargo

b) On board, using equal volumes from each tank

c) In a laboratory, in proportion to the volume in each tank

d) In a laboratory when S&W and API Gravity are the only tests needed

128. Which of the following types of sample container should be used to obtain a first foot sample?

a) Clear glass bottle

b) Amber glass bottle

c) Plastic bottle

d) Epoxy lined metal can

129. Which of the following types of container closure devices should not be used with glass bottles?

a) Plastic screw caps

b) Rubber stoppers

c) Metal screw caps

d) Cork stoppers

130. The definition of free water is:

a) The volume of water present in the tank that is not suspended in the oil

b) Any water found on the bob with water paste

c) Any water found using the M.M.C.

d) Any water that is trim corrected

131. A tank has a measured gauge height of 45' and is filled to a 40' innage with no free water. To take a middle spot sample, you must lower the sampler into the tank_____

a) 20' below the reference gauge point

b) 22.5' below the reference gauge point

c) 25' below the reference gauge point

d) 25' from the tank bottom

132. A tank has a measured gauge height of 45' and is filled to a 30' innage with no free water. To take a lower spot sample, you must lower the sampler into the tank____

a) 10' below the reference gauge point

b) 15’ above the tank bottom

c) 30’ below the reference gauge point

d) 40’ below the reference gauge point

133. A tank has a measured gauge height of 45’ and is filled to a 36’ innage with no free water. To take an upper spot sample, you must lower the sampler into the tank?

a) 12 feet above the tank bottom

b) 15 feet below the reference gauge point

c) 21 feet below the reference gauge point

d) 15 feet above the tank bottom

134. A tank has a measured gauge height of 45’ and is filled to a 40’ innage with no free water. To take a top spot sample you must lower the sampler into the tank_____

a) 5’ 00” below the reference gauge point

b) 5’ 06” below the reference gauge point

c) 11’ 08” below the reference gauge point

d) 12’ 06” below the reference gauge point

135. A floating roof tank is___

a) A tank that floats on its roof

b) A tank in which the roof floats freely on the surface of the liquid contents except at low levels when the weight of the roof is supported by its legs

c) A tank in which the roof, supported by guide wires, can be adjusted to the required height for safe filling of the tank

d) None of the above

136. The Total Calculated Volume is equal to the Gross Standard Volume plus____

a) Free water

b) S&W

c) Roof Correction

d) Free Water and S&W

137. For a trim correction to apply, which of the following conditions must exist?

a) Vessel must be down by the stern

b) Liquid may not contact the forward bulkhead

c) Liquid must touch all four bulkheads

d) All of the above

138. Roof corrections must be based on____

a) The observed API Gravity of the oil in the tank

b) The API Gravity at 60°F of the oil in the tank

c) The barrels per inch calculated from the tank capacity table

d) The critical zone

139. Ballast is____-

a) Water in the tanks of a vessel used for laundry and other sanitation purposes

b) Any water on board a vessel in any tank

c) Water that is used to clean cargo tanks

d) Water that allows the vessel to maintain stability and control stress and trim

140. Ballast is used for the purpose of____

a) Keeping the cargo warm

b) Segregation of the cargo

c) Reducing the ship's fuel consumption

d) Maintaining the vessel's stability, trim, and controlling vessel stress

141. A permanent ballast tank is____

a) A tank that contains ballast at all times

b) A tank that is designated to contain only ballast

c) A tank that is used only to maintain a permanent list condition

d) None of the above

142. A datum plate is___

a) A level metal plate located directly under the reference gauge point to provide a fixed contact surface from which liquid depth measurement can be made

b) A metal plate located next to the gauging point on a tank indicating the reference gauge height

c) A metal plate located close to the gauging point on a tank listing all the relevant tank data

d) A level metal plate located at the top of a gauge hatch on a tank from which the gauge height is measured

143. Deadwood is____

a) A wooden cup-case thermometer cases that are no longer fit for use

b) Any piece of gauging equipment made of wood (i.e. wooden handles of gauge tapes, cup-case thermometer cases) that have been exposed to chemicals and have been damaged as a result

c) Any tank fitting or structural member inside a tank that affects the capacity of the tank

d) None of the above

144. Total Observed Volume (TOV) is____

a) The total measured volume of all petroleum liquids, sediment and water, and free water at observed temperature

b) The total measured volume of all petroleum liquids, sediment and water but excluding free water, at observed temperature

c) The total volume of all petroleum liquids and sediment and water, corrected by the appropriate volume correction factor, for the observed temperature and API Gravity, to a standard temperature

d) The total measured volume of all petroleum liquids excluding water and sediment, at observed temperature

145. Gross Observed Volume (GOV) is____

a) The total measured volume of all petroleum liquids, sediment and water, and water at observed temperature

b) The total volume of all petroleum liquids and sediment and water, excluding free water, at observed temperature

c) The total volume of all petroleum liquids excluding sediment and water and free water, at observed temperature

d) The total volume of all petroleum liquids and sediment and water, excluding free water, corrected by the appropriate volume correction factor for the observed temperature and API Gravity, to a standard temperature

146. Gross Standard Volume (GSV) is____

a) The total volume of all petroleum liquids excluding free water, water and sediment, corrected by the appropriate volume correction factor for the observed temperature and API Gravity, to a standard temperature

b) The total volume of al petroleum liquids including free water, water and sediment, corrected by the appropriate volume correction factor for the observed temperature and API Gravity, to a standard temperature

c) The total volume of all petroleum liquids and sediment and water, excluding free water, corrected by the appropriate volume correction factor for the observed temperature and API Gravity, to a standard temperature

d) The total volume of all petroleum liquids including free water but excluding sediment and water, corrected by the appropriate volume correction factor for the observed temperature and API Gravity, to a standard temperature

147. Net Standard Volume (NSV) is____

a) The total volume of all petroleum liquids, excluding sediment and water and free water, corrected by the appropriate volume correction factor for the observed temperature and API Gravity, to a standard temperature

b) The total volume of all petroleum liquids, excluding sediment and water, but including free water, corrected by the appropriate volume correction factor for the observed temperature and API Gravity, to a standard temperature

c) The total volume of all petroleum liquids and free water, excluding sediment and water, corrected by the appropriate volume correction factor for the observed temperature and API Gravity, to a standard temperature

d) The total volume of all petroleum liquids and sediment and water and free water, corrected by the appropriate volume correction factor for the observed temperature and API Gravity, to a standard temperature

148. Total Calculated Volume (TCV) is____

a) GSV plus free water

b) NSV plus free water

c) GSV less sediment and water

d) NSV plus sediment and water

149. The total volume of everything in a tank at the observed temperature is called____

a) Total Calculated Volume (TCV)

b) Total Observed Volume (TOV)

c) On Board Quantity (OBQ)

d) Gross Observed Volume (GOV)

150. The volume of all material in a tank at the observed temperature less the free water is called__

a) Total Observed Volume (TOV)

b) Gross Standard Volume (GSV)

c) Gross Observed Volume (GOV)

d) Remaining On Board (ROB)

151. The volume of all material in a tank, less the free water, corrected by the volume correction factor is called____

a) Gross Standard Volume (GSV)

b) Gross Observed Volume (GOV)

c) Total Calculated Volume (TCV)

d) Net Standard Volume (NSV)

152. The reference gauge height of a tank is the distance from the___

a) Tank top to the tank bottom

b) Ullage hatch to the datum plate

c) Reference gauge point to the tank bottom or datum plate as documented

d) Tank bottom to the ullage hatch

153. An innage gauge measures___

a) The depth of the empty space above the liquid in a tank

b) The depth of the sediment in a tank

c) The length of an innage tape

d) The depth of the liquid in a tank

154. An ullage gauge measures____

a) The depth of the empty space above the liquid in a tank

b) The height of the free water in a tank

c) The length of an ullage tape

d) The depth of the liquid in a tank

155. The Total Observed Volume (TOV) is defined as___

a) The volume read from the strapping table

b) The total measured volume of all petroleum liquids, sediment and water, and free water at observed temperature and pressure

c) The volume read from the strapping table corrected for roof displacement

d) The volume used to calculate a Vessel Experience Factor (VEF)

156. The Gross Observed Volume (GOV) is defined is___

a) The volume read from the strapping table

b) The total volume of all petroleum liquids and sediment and water, excluding free water, at observed temperature and pressure

c) The volume read from the strapping table corrected for roof displacement

d) The volume used to calculate vessel ratios

157. The term 'Load on Top" is____

a) The shipboard practice of collecting water and settling water and oil mixtures resulting from ballasting and tank cleaning operations (usually in a slop tank) and subsequently loading cargo on top of and pumping the mixture ashore at the discharge port

b) Loading oil on to the deck of the vessel

c) Both of the above

d) Neither of the above

158. A Wall Wash test is____

a) The activity of rinsing a tank wall with a solvent to determine its compatibility with the product to be placed into the tank

b) The activity of rinsing a tank with clear, fresh water following tank cleaning to ensure its compatibility with the product to be placed into the tank

c) The activity of washing the walls of a tank to remove all traces of the product previously contained in the tank

d) All of the above

159. A shore pipeline is partially full before discharge and completely full after discharge. This will result in___

a) A gain of product as measured in the shore tank

b) A loss of product as measured in the shore tank

c) A loss of product as measured on the vessel

d) It will have no impact on the outturn

160. Factors that contribute to high evaporative losses are___

a) High vapor pressure of the cargo

b) Excessive agitation of cargo during voyage

c) Gauge hatches left open

d) All of the above

161. A Transit (or in-transit) Difference is____

a) The difference between the vessel measured volume at the loading port and the vessel measured volume at the discharge port

b) The difference between the vessel measured volume at the loading port and the shore measured volume at the loading port

c) The difference between the shore measured volume at the loading port and the shore measured volume at the discharge port

d) The difference between the shore line agreed tolerance and the loading port and the shore line agreed tolerance at the discharge port

162. If the vessel incurs a transit loss of product and a transit gain in water, you should___

a) Obtain samples of the free water

b) Check the vessel's bunkers and bunker consumption during the voyage

c) Verify the condition of seals on the sea suction and overboard discharge valves

d) All of the above

163. The preferred device for taking temperatures in a marine custody transfer is____

a) A mercury-in-glass thermometer in a cupcase assembly

b) An in-line temperature probe

c) An on board S.A.A.B. System

d) A portable electronic thermometer

164. The term "simultaneous ballasting or deballasting" means___

a) The vessel is transferring ballast from one ballast tank to another

b) The vessel is taking on or pumping off ballast in more than one tank at a time

c) The vessel is transferring ballast while cargo is being pumped

d) The Chief Officer has been authorized to pump ballast a shore

165. Measurement of free water on board marine vessels is important because___

a) Free water volumes determined from vessel measurements are often deducted from shore receipt volumes after discharge

b) U. S. Customs Service requires it on imported cargoes

c) An accurate Voyage Analysis Report cannot be completed without accurate vessel free water measurements

d) All of the above

166. An OBQ inspection is performed___

a) Before loading a clean product

b) Before any cargo is loaded

c) Before any chemical is loaded

d) Before loading a clean product after a dirty one

167. If the ROB is non-liquid in nature what is the preferred measurement method?

a) A single innage

b) An average of multiple innages

c) By ullage

d) Using trim corrections

168. If a series of innage gauges indicates that the ROB / OBQ lies evenly across the bottom, you should determine the volume by____

a) Use of trim corrections

b) Applying the wedge formula

c) By using an average of the innage gauges.

d) By using the innage at the official gauge point

169. When non-liquid ROB is measured, it is considered to be evenly distributed across the tank bottom except___

a) When the vessel is listing

b) When the cargo was heated

c) When multiple gauges in the tank prove otherwise

d) When ROB is more than 4" deep

170. The amount of material found in a tank prior to loading is known as___

a) Slops

b) On Board Quantity [OBQ]

c) Bunkers

d) Remaining On Board [ROB]

171. The amount of material found in a tank after discharge is known as___

a) Slops

b) On Board Quantity [OBQ]

c) Bunkers

d) Remaining On Board [ROB]

172. When portable electronic gauging equipment is used on board a marine vessel, which of the following considerations needs to be addressed?

a) The equipment used must securely fit the vapor control valve

b) The vessel's tank capacity tables must have been adjusted to accommodate the vapor control valve location and reference height

c) The equipment should be grounded

d) All of the above

173. The Reference Gauge Height of a vessel tank is___

a) The overall height of the expansion trunk, referred to in the drawings

b) The distance from the tank bottom to the reference gauge point as specified on the tank's capacity table

c) The measured distance from the tank bottom to the reference gauge point

d) The place inside the tank where automatic measurement floats are installed

174. How many liquid level measurements must be taken in a vessel's tanks when the vessel is in motion (rolling)?

a) 2

b) 3

c) 4

d) 5

175. When a ship is trimmed by the head?

a) The aft draft reading is greater that the forward draft reading

b) The forward draft reading is greater than the aft draft reading

c) The ship has water in the forepeak tank

d) Trim corrections will always be added to the measured gauge

176. When a ship is trimmed by the stern?

a) The aft draft reading is greater that the forward draft reading

b) The forward draft reading is greater than the aft draft reading

c) The ship has water in the aft peak tank

d) Trim corrections will always be subtracted from the measured gauge

177. Trim corrections are applicable to:

a) Only the ROB quantity

b) Only the OBQ quantity

c) Any non-liquid material

d) Any liquid material that is touching all four tank bulkheads

178. When a vessel is not on an even keel, tank gauges must be corrected by the use of___

a) Volume correction tables or volume correction calculations

b) Trim correction tables or trim calculations

c) Weight correction tables or weight correction calculations

d) Draft correction tables or draft correction calculations

179. A wedge condition will exist if the liquid in a tank____

a) Covers the bottom of the tank.

b) Touches three of the four bulkheads

c) Accumulates beneath the gauge hatch

d) All of the above

180. Which of the following conditions must be present for trim corrections to apply?

a) Tank contents must touch all four bulkheads

b) Tank contents must be non liquid

c) Tank contents must not contact the forward bulkhead

d) (a) and (b)

181. A wedge formula calculation on a vessel trimmed by the stern may be used when___

a) Liquid material does not contact the forward bulkhead

b) Solid material is gauged at a single gauge point

c) Free water completely covers the tank bottom

d) All of the above

182. Freeboard on a vessel is___

a) The distance from the waterline to the vessel's deck level

b) The distance from the waterline to the vessel's keel

c) The time when U. S. Customs Service permits others to board the vessel

d) The time of day that lay time begins according to the Charter Party

183. Draft readings can be used to obtain ____

a) The depth of the vessel in the water

b) The trim of the vessel

c) The list of the vessel

d) All of the above

184. A pre-loading tank inspection key meeting between vessel personnel, shore personnel and inspection personnel should determine___

a) Tank Number, tank capacity, intended cargo volume

b) The last three cargoes and method of tank cleaning

c) The contents of adjacent tanks

d) All of the above

185. When reading the draft marks on a barge or vessel, which part of the number indicates the actual zero point (start) of the number in question?

a) The lower edge of the number

b) The upper edge of the number

c) The mid point of the number

d) None of the above

186. The **critical zone** in a shore tank defines that part of the tank where___

a) The flexing bottom of a tank is at its highest point

b) The point where the tank will overflow if more liquid is added

c) The floating roof is no longer resting on its legs

d) The vertical section of a tank identified on the capacity table where the floating roof is only partially afloat and the table may be inaccurate

187. A Master Tape is____

a) The only tape that may be used to take custody transfer measurements

b) A tape belonging to the Captain of the vessel

c) A tape to be used only for tank calibration

d) A reference tape, traceable to a recognized international standard, which is to be used only for verifying the calibration of tapes used in the field

188. An innage gauge measures____

a) The distance from the surface of the liquid in the tank to the reference gauge point of the tank

b) The distance between the point where the floating roof of the tank is floating freely and the point where it is resting fully on its supports

c) The level of liquid in a tank measured form the datum plate or tank bottom to the surface of the liquid

d) The distance from the tank datum plate or bottom to the tank reference gauge point

189. An outage gauge measures____

a) The distance from the tank datum plate or bottom to the tank reference point

b) The distance from the surface of a liquid in a tank to the reference gauge point of the tank

c) The amount of product transferred out from a tank

d) The level of liquid in a tank measured from the tank bottom to the surface of the liquid

190. The observed reference height of a tank is___

a) The distance from the reference point to the bottom of the tank or the datum plate as measured during the gauging operation

b) The distance from the reference point to the bottom of the tank or the datum plate as shown on the tank capacity tables

c) The distance from the ullage pipe to the liquid level

d) Usually written somewhere on the ullage pipe

191. Innage is best described by_____

a) The distance from the datum plate or the tank bottom to the surface of the product

b) The measurement from the tank bottom to reference point

c) The cut found on the bob

d) The distance from the reference point to surface of the product

192. An outage (or ullage) gauging bob is designed to be used with____
   a) An outage (or ullage) gauging tape
   b) An innage tape
   c) Can be used with either
   d) A sounding rod

193. The zero point of an outage gauging bob is located at the_____
   a) The tip of the bob
   b) The mid point of the bob
   c) The inside top of the eye of the bob
   d) The top of the tape swivel

194. An extension outage (or ullage) bob is designed to be used with_________
   a) An outage (or ullage) gauging tape
   b) An innage tape
   c) Can be used with either
   d) A sounding rod

195. The distance between the point where the floating roof begins to rest on its normal supports and the point where it begins to float freely, is known as______
   a) The floating level
   b) The lower leg level
   c) The critical zone
   d) The displacement level

196. On a shore tank, the distance between the Reference Gauge Point and the Datum Point, as measured at the time of gauging, is called_______
   a) Observed reference point
   b) Reference gauge height
   c) Total gauge height
   d) Observed gauge height

197. A tank datum plate is____
   a) The position on a tank where the gauge height is noted
   b) The point marked on the gauge hatch of a tank to indicate the position from which the tank is to be gauged
   c) The plate on the tank shell that lists the tank general data such as roof weight, height of tank, etc
   d) A plate placed in the tank and directly under the reference gauge point to provide a fixed contact surface

198. When rainwater or snow accumulates on the roof of an external floating roof tank, the product level in the tank will______
   a) Decrease
   b) Remain the same
   c) Increase
   d) None of the above

199. Which of the following statement is true?
   a) An innage tape and bob may be used to take an outage gauge.
   b) An ullage gauge is the same thing as an outage gauge.
   c) The use of unslotted standpipes is not recommended for custody transfer measurements.
   d) All of the above

200. The measured distance from the datum plate or tank bottom to the reference gauge point is called ___

a) Observed gauge height
b) Innage gauge height
c) Reference gauge height
d) Ullage gauge height

201. The distance from the datum plate or tank bottom to the reference gauge point, as shown on the tank capacity table is called ___

a) Observed gauge height
b) Innage gauge height
c) Reference gauge height
d) Ullage gauge height

202. The accuracy of tank's capacity table can be affected if____

a) The tank bottom flexes with the transfer of contents
b) Expansion at the vertical middle (barreling) occurs during filling
c) The tank shell has accumulated deposits of previous contents
d) All of the above

203. If an electronic gauging tape (PMU) is used to measure free water, which of the following procedures should be followed?

a) Rely on the PMU alone for free water level measurement because that is the same instrument used to measure the oil level
b) Apply water-finding paste to the bob of the PMU and compare the water level indicated by the paste to the water level indicated by the PMU
c) Rely on water-finding paste alone
d) Recommend to the terminal that the free water received and measured ashore be applied to the vessel figures

204. To determine the ullage of the liquid in a tank using an innage tape and bob assembly, you should____

a) Read the immersion depth of the tape at the reference gauge point and subtract the oil cut reading on the bob
b) Read the immersion depth of the tape at the reference gauge point and add the oil cut reading on the bob
c) Subtract the oil cut on the bob from the tank reference gauge height indicated on the tank capacity table
d) Subtract the oil cut on the bob from the measured tank reference gauge height

205. When using water indicating paste in heavy oils, which of the following actions should you take to enable the paste to be read more easily?

a) Gently blowing on the gauging bar to remove the heavy oil
b) Use a soft cotton rag or towel to remove the excess oil
c) Use a suitable solvent to gently wash to surface of the paste

d) Immerse the gauging bar in a container full of light product such as gasoline

206. A quantity of free water (water bottoms) may be maintained within a store tank for which of the following reasons?

a) To permit easy determination of the shore tank free water quantity

b) So that a certain percentage of free water can be pumped with each movement in order to aid in S&W blending of cargoes

c) To negate any effect that diaphragming (bottom flexing) of the tank bottom would have on tank measured quantities

d) To aid in detection of any product seepage from the tank

207. Gauging standpipes on floating roof tanks should be___

a) At least 8 inches in diameter

b) Extend to within 12 inches of the tank bottom

c) Have two rows of overlapping slots located on opposite sides of the pipe

d) All of the above

208. When gauging a tank by the innage method, a comparison between the observed and reference gauge height is made to ensure which of the following?

a) The gauge tape and bob are suspended in a vertical position within the tank

b) The gauge bob is in contact with the tank bottom or datum plate

c) The tape has not been lowered too far into the tank

d) All of the above

209. If the density of a product within a tank is greater than the density of water, where would you expect to find any free water located in this tank?

a) You would not find it as it would not settle out of this product but would remain in suspension

b) On top of the product

c) Below the product

d) None of the above

210. Reference heights should be____

a) Recorded from the calibration tables, before you start the tank inspection

b) Checked against the observed reference height

c) Included in the inspection report

d) All of the above

211. Free water measurements may be taken by_____

a) The innage method

b) The ullage method

c) The Heimlich method

d) Both (a) and (b)

212. When would you check for water on the top of a cargo?

a) If the observed API Gravity is greater than 10.0

b) If the observed API Gravity is less than 10.0

c) You never check for the presence of water on top of a cargo

d) When you are in the Southern Hemisphere

213. Which of the following is acceptable for the determination of free water for marine vessel custody transfer measurements?

a) A portable electronic gauge tape

b) A bob with water-finding paste

c) None of the above

d) Both (a) and (b)

214. The floating roof of a shore tank displaces a certain volume of liquid when it is in the free floating position. The weight of the volume of liquid displaced is___

a) Equal to the weight of the roof and the attached deadwood

b) Greater than the weight of the roof and the attached deadwood

c) Less than the weight of the roof and the attached deadwood

d) None of the above

215. What is the purpose of a floating roof adjustment in a shore tank calculation?

a) To account for the compression on the liquid due to the weight of the roof

b) To account for the volume of liquid displaced as a result of the weight of the roof

c) The temperature of the roof affects the temperature of the petroleum liquid in the tank

d) The roof thickness changes with the API of the petroleum liquid

216. If a tank has more than ten feet of liquid, what is the minimum number of temperature readings that must be taken?

a) 5

b) 3

c) 1

d) One every 2 feet

217. If only one temperature is required where this temperature should be taken from?

a) The middle of the upper third

b) The middle of the liquid

c) The middle of the lower third

d) Use a side readout thermometer

218. The quickest way to stabilize the reading from a thermometer is to____

a) Allow the probe to stay in the product for twice the required time

b) Move the probe up and down at least one foot above and below the spot the temperatures is to be taken

c) Use fresh batteries

d) There is no way to speed up the temperature process

219. What is the minimum amount of product needed for a temperature to be taken?

a) Whenever there is sufficient material present to immerse the probe

b) One foot

c) Ten feet

d) Only when the roof is free floating

220. A portable electronic thermometer should be read and recorded to the nearest____

a) Half degree

b) One degree

c) One tenth of a degree

d) None of above

221. What is the minimum number of temperatures to be taken on a marine vessel with tanks containing more than 5000 Bbls?

a) Three per tank

b) One per tank

c) Weighted average per tank

d) None of above

222. "In motion" is defined as continuously raising and lowering the probe above and below the desired temperature measurement depth, by approximately how much?

a) Six inches

b) One inch

c) One foot

d) Three feet

223. The term "horizontal temperature stratification" means____

a) The temperature in a tank is different near the tank center than at the gauge hatch

b) Any difference in temperature measured at different levels in a tank

c) Only if the difference measured at any two levels in a tank exceeds 5°F

d) All of the above

224. What is the minimum number of temperatures to be taken in a tank containing more than 10 feet of product?

a) 1

b) 3

c) 5

d) 2

225. What temperature is to be used during an ROB survey if there is only 2 inches (5cm) of liquid in the bottom of the tank?

a) The measured temperature at the middle of the liquid

b) 60°F

c) The average temperature of the product in the tank before it was discharged

d) The temperature stated by the vessel's representative

226. A 25,000 barrel capacity storage tank has a depth of product of 12 feet in the tank. What is the minimum number of temperature measurements that should be obtained on this tank?

a) One

b) Two

c) Three

d) Four

227. A pipeline carries water at 8 ft/s through 1 ft ID pipe. At another section the inside diameter is 2 ft. What is the velocity at this section?

a) 2 ft/s

b) 16 ft/s

c) 4 ft/s

d) 32 ft/s

228. A pump is needed to move a high viscosity, heavy (high specific gravity) crude oil from a storage tank to a treating facility for processing. The tank is not insulated and it is located in a cold-weather environment. The field from which this oil is produced has a low reservoir pressure, and the oil is produced in slugs that re-fill the tank sporadically. The oil must be moved at a closely controlled rate to the treating facility to avoid overloading it. What type pump, based on service capabilities and cost, should be recommended for this service?

a) Centrifugal pump

b) Gear pump

c) Reciprocating (plunger) pump

d) Turbine pump

229. To restrict reversal of flow, ________ may be used

a) check valve

b) needle valve

c) globe valve

d) All of the above

230. Which of the following is true about pigging?

a) It is a process of pushing a device equipped with metal wire brushes in the pipeline

b) It is used to clean the deposits on the inner walls of the pipeline

c) Progressive pigging is a special method for pigging lines with reduced internal diameters

d) All of the above

231. Which of the following formula is used to calculate the shell thickness of a cylindrical pressure vessel?

a) $t = \frac{pd}{2fj}$ where, t is thickness of shell, d is the diameter of the cylinder, p is the design pressure, f is the allowable stress and j is the joint efficiency

b) $t = \frac{pd}{4fj}$ where, t is thickness of shell, d is the diameter of the cylinder, p is the design pressure, f is the allowable stress and j is the joint efficiency

c) $t = \frac{pd}{3fj}$ where, t is thickness of shell, d is the diameter of the cylinder, p is the design pressure, f is the allowable stress and j is the joint efficiency

d) $t = \frac{pd}{5fj}$ where, t is thickness of shell, d is the diameter of the cylinder, p is the design pressure, f is the allowable stress and j is the joint efficiency

232. Which of the following formula is used to calculate the shell thickness of a spherical pressure vessel?

a) $t = \frac{pd}{2fj}$ where, t is thickness of shell, d is the diameter of the cylinder, p is the design pressure, f is the allowable stress and j is the joint efficiency

b) $t = \frac{pd}{4fj}$ where, t is thickness of shell, d is the diameter of the cylinder, p is the design pressure, f is the allowable stress and j is the joint efficiency

c) $t = \frac{pd}{3fj}$ where, t is thickness of shell, d is the diameter of the cylinder, p is the design pressure, f is the allowable stress and j is the joint efficiency

d) $t = \frac{pd}{5fj}$ where, t is thickness of shell, d is the diameter of the cylinder, p is the design pressure, f is the allowable stress and j is the joint efficiency

233. Which of the following sentence is false about the parallel flow heat exchangers?

a) In parallel flow heat exchanger, the outlet temperature of cold fluid my exceed that of the cold fluid

b) Parallel flow is used for cold viscous fluids

c) Both (a) and (b)

d) None of the above

234. Which of the following sentence is false about the counter flow heat exchangers?

a) In counter flow heat exchanger, the outlet temperature of cold fluid my exceed that of the cold fluid

b) In a pure counter flow heat exchanger, the temperature difference between the hot and cold fluids is not constant throughout.

c) Both (a) and (b)

d) None of the above

235. In a heat exchanger, floating head is provided to

a) Facilitate the cleaning of the exchanger

b) Increase the heat transfer area

c) Relieve stresses caused by thermal expansion

d) Increase the log mean temperature gradient

236. The advantage of using a 1-2 shell and tube heat exchanger over a 1-1 shell and tube heat exchanger is

a) Lower tube side pressure drop
b) Lower shell side pressure drop
c) Higher tube side heat transfer coefficient
d) Higher shell side heat transfer coefficient

237. In a finned tube heat exchanger:
a) Only heat transfer area is augmented
b) Only film coefficient is augmented
c) Both heat transfer area and film coefficient are augmented
d) None of the above

238. The log mean temperature difference for heat exchanger is given by:
a) $LMTD = \frac{\Delta T_o - \Delta T_L}{ln\left(\frac{\Delta T_o}{\Delta T_L}\right)}$, where $\Delta T_o = T_{h,out} - T_{c,in}$ and $\Delta T_L = T_{h,in} - T_{c,out}$ for counter flow
b) $LMTD = \frac{\Delta T_o - \Delta T_L}{ln\left(\frac{\Delta T_L}{\Delta T_o}\right)}$, where $\Delta T_o = T_{h,out} - T_{c,in}$ and $\Delta T_L = T_{h,in} - T_{c,out}$ for counter flow
c) $LMTD = \frac{\Delta T_L - \Delta T_o}{ln\left(\frac{\Delta T_o}{\Delta T_L}\right)}$, where $\Delta T_o = T_{h,out} - T_{c,in}$ and $\Delta T_L = T_{h,in} - T_{c,out}$ for counter flow
d) None of these

239. The true temperature difference for multipass and cross flow heat exchangers is given by $\Delta T_{corrected} = FT \times LMTD$ for counter flow. Which of the following statement is false about FT?
a) It is unity for true conterflow heat exchanger
b) It is less than unity for multipass arrangements
c) It is greater than unity for multipass arrangement
d) None of the above

240. The following informations are given for the sizing of a two phase vertical separator

Flow rate 10 MMScfd at 0.6 sp. Gr.

2000 bopd at 40°API

Operating pressure: 1000 psia

Temperature: 60°F

Compressibility Factor = 0.84

Liquid droplet size: 140 micron

Drag Coefficient: 0.851

The internal diameter of the separator will be
a) 21.8
b) 32
c) 36
d) 42

241. The following information are given for the sizing of a heater treater:

Oil Flow rate 2000 bpd

Oil gravity 40°API

Inlet oil temperature 90°F

Outlet Temperature: 120°F

Water S.G. 1.04

Inlet BS & W: 10%

Outlet BS&W: 1%

The amount of heat required will be:

a) 0.49 MMBtu/h

b) 0.16 MMBtu/h

c) 0.50 MMBtu/h

d) 1.72 MM Btu/h

242. Three storage tanks are delivered to the site of a new producing well. The tanks appear to be unusual in size. No tank table was delivered with the tanks. The well is to be placed on production as soon as the tanks are set. The basic dimensions of the tanks are:

Tank outside diameter: 15.0 ft

Shell thickness: 0.35-in.

Tank shell height: 17 ft

Overflow line outlet: base, 8-in. from top of tank

What is the usable storage capacity of the tank battery to the nearest barrel?

a) 509

b) 1,527

c) 1,539

d) 1,593

243. A platform is producing surplus gas that must be sold into a nearby pipeline; flaring is not permitted. The gas has the following characteristics:

Gas gravity = 0.689 SG (Air = 1)

Ratio of specific heats = 1.206

Molecular weight = 20

Super compressibility factor = 0.022

Wells on the platform produce a total of 18 mmcfd dry gas (after separation and dehydration). Platform fuel use is 6 mmcfd. The gas is cooled after leaving the dehydrator to a temperature of 60°F. Pressure at the gas cooler outlet is 227 psig. Pipeline pressure is 1,050 psig. What minimum horsepower, hp, would be required to compress the surplus gas from this platform for sale into the pipeline?

a) 1,191

b) 1,238

c) 1,787

d) 2,649

Statement for Linked Answer Questions for 244-245

For a 1-1 fixed tubes heat exchanger, the following data are given:

Design inside pressure Shell Side 10^3 KN/m^2

Tube Side 2×10^3 KN/m^2

Design Temperature

Shell Side 50°C

Tube Side 150°C

Allowable stress value at both design temperatures is 1.2×10^5 KN/m^2. Shell inside diameter = 0.8 m, Corrosion allowance (for shell and headers) = 3×10^{-3} m and joint efficiency = 1

244. The thickness of the shell will be:

a) 5.48×10^{-3} m

b) 4.48×10^{-3} m

c) 3.48×10^{-3} m

d) 2.48×10^{-3} m

245. The header shell thickness will be

a) 0.01948 m

b) 0.02948 m

c) 0.03948 m

d) 0.04948 m

Statement for Linked Answer Questions for 246-248

An organic liquid is cooled from 105°C to 50°C in a one shell pass and two tube pass heat exchanger by circulating water and other details are given below:

Cooling water inlet temperature = 25°C

Cooling water outlet temperature = 40°C

Heat capacity for water = 4.2 KJ/Kg °C

Heat capacity for hot liquid = 2.84 KJ/kg. °C

Overall heat transfer coefficient = 600 w/m^2.°C

FT = 0.85

246. The value of LMTD will be

a) 44.86°C

b) 51.86°C

c) 41.86°C

d) 21.86°C

247. The value of true temperature difference will be:

a) 35.58°C

b) 25.58°C

c) 15.58°C

d) 25.58°C

248. The heat transfer area for this exchanger will be

a) 10.322 m^2

b) 20.322 m^2

c) 30.322 m^2

d) 40.322 m^2

249. A shell and tube heat exchanger is constructed using single pass horizontal tubes to condense the steam at 54°C on the outside of the tube. The cooling water enters at 20°C and leaves at 36°C. What will be the value of log mean temperature difference?

a) 15.16°C

b) 35.16°C

c) 45.16°C

d) 25.16°C

250. The seam-to-seam length (ft) of a two-phase 30-inch-I.D. horizontal separator needed to handle 1,500 bbl/d and 500 Mscf/d with a liquid retention rate of 3 minutes and operating fluid level at 50% is most nearly:

a) 4.9

b) 7.1

c) 9.6

d) 10.6

Answers									
Q1.	b	Q36.	c	Q71.	b	Q106.	c	Q141.	b
Q2.	d	Q37.	a	Q72.	b	Q107.	c	Q142.	a
Q3.	a	Q38.	c	Q73.	a	Q108.	c	Q143.	c
Q4.	b	Q39.	c	Q74.	c	Q109.	d	Q144.	a
Q5.	a	Q40.	a	Q75.	c	Q110.	c	Q145.	b
Q6.	a	Q41.	d	Q76.	c	Q111.	b	Q146.	c
Q7.	a	Q42.	c	Q77.	c	Q112.	a	Q147.	a
Q8.	a	Q43.	d	Q78.	a	Q113.	c	Q148.	a
Q9.	d	Q44.	a	Q79.	a	Q114.	c	Q149.	b
Q10.	c	Q45.	a	Q80.	b	Q115.	c	Q150.	c
Q11.	c	Q46.	c	Q81.	c	Q116.	c	Q151.	a
Q12.	c	Q47.	a	Q82.	c	Q117.	c	Q152.	c
Q13.	c	Q48.	b	Q83.	a	Q118.	a	Q153.	d
Q14.	a	Q49.	a	Q84.	c	Q119.	b	Q154.	a
Q15.	c	Q50.	b	Q85.	c	Q120.	a	Q155.	b
Q16.	d	Q51.	a	Q86.	b	Q121.	d	Q156.	b
Q17.	d	Q52.	d	Q87.	d	Q122.	c	Q157.	a
Q18.	d	Q53.	b	Q88.	d	Q123.	b	Q158.	a
Q19.	c	Q54.	a	Q89.	b	Q124.	b	Q159.	b
Q20.	a	Q55.	a	Q90.	b	Q125.	b	Q160.	d
Q21.	a	Q56.	b	Q91.	c	Q126.	c	Q161.	a
Q22.	d	Q57.	a	Q92.	a	Q127.	c	Q162.	d
Q23.	d	Q58.	a	Q93.	d	Q128.	a	Q163.	d
Q24.	b	Q59.	a	Q94.	b	Q129.	b	Q164.	c
Q25.	a	Q60.	c	Q95.	c	Q130.	a	Q165.	d
Q26.	a	Q61.	c	Q96.	b	Q131.	c	Q166.	b
Q27.	d	Q62.	a	Q97.	d	Q132.	d	Q167.	b
Q28.	a	Q63.	c	Q98.	c	Q133.	b	Q168.	c
Q29.	a	Q64.	a	Q99.	c	Q134.	b	Q169.	c
Q30.	c	Q65.	a	Q100.	d	Q135.	b	Q170.	b
Q31.	a	Q66.	a	Q101.	d	Q136.	a	Q171.	d
Q32.	c	Q67.	b	Q102.	c	Q137.	c	Q172.	d
Q33.	c	Q68.	c	Q103.	a	Q138.	a	Q173.	b
Q34.	d	Q69.	a	Q104.	a	Q139.	d	Q174.	d
Q35.	c	Q70.	c	Q105.	b	Q140.	d	Q175.	b

Answers

Q176.	a	Q191.	a	Q206.	c	Q221.	a	Q236.	c
Q177.	d	Q192.	a	Q207.	d	Q222.	c	Q237.	a
Q178.	b	Q193.	c	Q208.	d	Q223.	b	Q238.	a
Q179.	b	Q194.	a	Q209.	b	Q224.	b	Q239.	c
Q180.	a	Q195.	c	Q210.	d	Q225.	b	Q240.	a
Q181.	a	Q196.	d	Q211.	d	Q226.	c	Q241.	a
Q182.	a	Q197.	d	Q212.	b	Q227.	a	Q242.	b
Q183.	d	Q198.	c	Q213.	d	Q228.	b	Q243.	a
Q184.	d	Q199.	d	Q214.	a	Q229.	a	Q244.	a
Q185.	a	Q200.	a	Q215.	b	Q230.	d	Q245.	a
Q186.	d	Q201.	c	Q216.	b	Q231.	a	Q246.	c
Q187.	d	Q202.	d	Q217.	b	Q232.	b	Q247.	a
Q188.	c	Q203.	b	Q218.	b	Q233.	a	Q248.	b
Q189.	b	Q204.	a	Q219.	a	Q234.	b	Q249.	d
Q190.	a	Q205.	c	Q220.	c	Q235.	c	Q250.	c

CHAPTER 6

Offshore Drilling and Production Practices

1. Which of the following is an example of bottom supported mobile offshore drilling unit?
   a) Semi-Submersible Rig
   b) Drill Ship
   c) Jack-Up Rig
   d) None of the above

2. Which of the following is an example of floating drilling rig?
   a) Drill Barge
   b) Jack-Up Rig
   c) Drill Ship
   d) None of the above

3. Which of the following rig is used for deep water?
   a) Drill Barge
   b) Jack-Up Rig
   c) Drill Ship
   d) None of the above

4. Which of the following is true in case of Semi-Submersible Rigs?
   a) They can be moved from one location to other location
   b) They partially flood their pontoons to achieve desired height above the water
   c) They may be held in place by mooring lines attached to the seafloor anchor
   d) All of the above

5. Which of the following can't be used for deep water?
   a) Drilling Barge
   b) Semi-Submersible Rigs
   c) Drillships
   d) None of the above

6. Which of the following is true in case of Drillship?
   a) It can be moved from location to location under their own power
   b) It uses dynamic positioning system for maintaining their position over the drilling location
   c) It has high storage capacity
   d) All of the above

7. Which of the following is not common design criteria used for selecting a proper marine rig:
   a) Water depth rating
   b) Manufacturer name
   c) Derrick and substructure capacity
   d) Deck load capacity

8. Which of the following is false about drillships?
   a) They are always anchored
   b) They often use dynamic positioning
   c) They are usually self-propelled for rig moves
   d) None of the above

9. Which of the following is false about jack up rigs?
   a) Usually they have 3 legs which stand on the seabed
   b) Hull is lowered and legs raised for rig moves
   c) They can be used to drill in deep waters (> 5000ft)
   d) BOP's are below the derrick cantilever
   e) None of the above

10. In offshore drilling, the BOP is:
   a) Installed on the wellhead located on the top of the riser pipe
   b) Installed on the wellhead located on the top of the casing at the sea bed
   c) Both (a) and (b)
   d) None of the above

11. Dynamic Positioning uses
   a) No Rope/Wire
   b) thrusters and propellers to keep the vessel in position
   c) Both (a) and (b)
   d) None of the above

12. Which of the following platform is anchored to the seabed and hold most of the equipments above the surface?
   a) Jack up rig
   b) Complaint towers
   c) Both (a) and (b)
   d) Drillship

13. A system connecting seabed facilities to surface facilities is called
   a) Riser
   b) Wellhead
   c) Accumulator
   d) BOP

14. A chain/rope system keeping floating surface facilities in place is called
   a) Mooring
   b) Wireline
   c) draworks
   d) None of the above

15. Which of the following is the function of riser?
   a) Guide fluid and tools for well drilling and workover
   b) Guide fluid from wells to floaters injection
   c) Guide fluid from floaters to well s export
   d) All of the above

16. Depth of riser for shallow water is
   a) < 500m
   b) 500-2000m
   c) >2000m
   d) None of the above

17. Depth of riser for Deep water is
   a) < 500m
   b) 500-2000m
   c) >2000m
   d) None of the above

18. Depth of riser for Ultra deep water is
   a) < 500m
   b) 500-2000m
   c) >2000m
   d) None of the above

19. Which of the following is an example of floating production system?
   a) FPSO(floating production, storage, and offloading system),
   b) FSO (floating storage and offloading system),
   c) FSU (floating storage unit).
   d) All of the above

20. TLPs are used in water depths up to about the depth of
   a) 6,000 feet
   b) 8000 feet
   c) 9500 feet
   d) 10500 feet

21. The __________ is used to maintain position of the vessel.
   a) Dynamic positioning system
   b) Global positioning system
   c) Both (a) and (b)
   d) None of the above

22. A Drillship can drill in water depths up to a depth of
   a) 12,000 feet
   b) 18000 feet
   c) 5000 feet
   d) 7500 feet

23. Compliant towers are designed to sustain significant
   a) lateral deflections and forces
   b) longitudinal deflections and forces
   c) Both (a) and (b)
   d) None of the above

24. FPSO system is preferred for
   a) Small onshore field
   b) Large offshore field
   c) Small offshore field
   d) Large onshore field

25. Compliant towers are typically used in water depths ranging from
   a) 1,500 and 3,000 feet (450 and 900 m)
   b) 2000 and 3000 feet (600 and 900 m)
   c) 3000 and 4000 feet (900 and 1300 m)
   d) None of the above

26. Fixed platforms are economically feasible for installation in water depths up to
   a) about 1,700 feet (520 m)
   b) about 2500 feet (800 m)
   c) about 3000 feet (900 m)
   d) about 4000 feet (1300 m)

27. The primary factor(s) for the design of offshore structure is(are)____

    a) Wind

    b) Wave

    c) Current

    d) All of the above

28. World's deepest spar is located in

    a) Gulf of Mexico

    b) Mumbai high

    c) Arabian sea

    d) None of the above

29. Which of the following are the main advantages of Drill ship:

    a) It is mobile with high speed transit

    b) Deck load and total load of drill ship is greater than jack-up and semisubmersible

    c) Low mobilization cost

    d) All of the above

30. Which of the following systems are required for offshore oil and gas production?

    a) Subsea manifold and connection systems

    b) Jumpers and Risers, Umbilical systems

    c) Subsea wellhead and Subsea tree

    d) All of the above

31. __________ is a bundled arrangement of tubing, piping, and/or electrical conductors in an armored sheath that is installed from the host facility to the subsea production system equipment.

    a) Subsea manifold and connection systems

    b) Jumpers

    c) Risers

    d) Umbilical systems

32. ____________ is used to transmit the control fluid and/or electrical current necessary to control the functions of the subsea production and safety equipment.

    a. Subsea manifold and connection systems

    b. Jumpers

    c. Risers

    d. Umbilical systems

33. Which of the following is used to specifically monitor pressures and inject fluids from the host facility to critical areas within the subsea production equipment?

    a) Umbilical systems

    b) Jumpers

    c) Risers

    d) None of the above

34. ____________ is the portion of the flowline that resides between the host facility and the seabed adjacent to a host facility.

    a) Risers

    b) Umbilical systems

    c) Jumpers

    d) None of the above

35. The length of the riser mainly depends on
    a) Water depth
    b) Riser configuration
    c) Both (a) and (b)
    d) None of the above

36. ___________ forms a conduit for transporting of the drill-string and drilling fluid to the hole bottom and protecting them from harsh subsea conditions.
    a) Drilling riser
    b) Jumper
    c) Umbilical systems
    d) None of the above

37. Which of the following is the example of floating offshore structure?
    a) Jack up rig
    b) Compliant tower
    c) SPAR platform
    d) None of the above

38. Which of the following rig is similar to drilling barge?
    a) Submersible rigs
    b) Jackup rigs
    c) Both (a) and (b)
    d) None of the above

39. In ______________, the drilling structure is supported by large submerged pontoons that are flooded and rest on the sea floor when drilling.
    a) Submersible rigs
    b) Tension Leg Platforms
    c) FPSO
    d) None of the above

40. Which types of platform are fixed structure but they use narrower and flexible tower of concrete and steel?
    a) Tension leg platform (TLP)
    b) Drillship
    c) Barge
    d) None of the above

41. __________ is designed to receive from nearby oil platform process it, store it and transfer it to the tankers for its further transportation to the ports?
    a) TLP
    b) Drillship
    c) Spar
    d) FPSO

42. Which of the following configuration of the risers are used in offshore oil and gas drilling & production system?
    a) Top tension risers
    b) Compliant risers
    c) Steel catenary risers
    d) All of the above

43. In ________, wellhead is connected to the surface directly via a riser conduit and production tubing hanger installed at the surface.
    a) Subsea completion
    b) Tie back completion
    c) Both (a) and (b)
    d) None of the above

44. In _______, subsea tree is connected to the surface via jumper, manifold, flow lines & import risers and production tubing hanger installed subsea to the tubing head.

a) Subsea completion
b) Tie back completion
c) Both (a) and (b)
d) None of the above

45. Which of the following riser is most widely used for ultra deep water wells?

a) Top tension risers
b) Compliant risers
c) Steel catenary risers
d) None of the above

46. Which of the following design criteria is used for an umbilical installation?

a) Minimum bending radius of umbilical
b) Maximum allowable tension and compression loads
c) Lateral stability on the sea bed
d) All of the above

47. Which of the following installation method is used for pipelines/risers?

a) S-Lay method
b) Z-lay method
c) C-laying method
d) None of the above

48. In ______, the pipe is laid form a near horizontal position on a lay barge using a combination of horizontal tension and a stinger.

a) S-lay mehod
b) J-lay method
c) Towing-lay method
d) None of the above

49. In _______, the pipe is laid from an elevated tower on a lay barge using longitudinal tension without an over bend configuration at the sea surface.

a) S-lay method
b) J-lay method
c) Towing-lay method
d) None of the above

50. In ______, pipe is made up at some remote location onshore, spooled onto a large radius reel abroad a reel lay vessel and then unreeled, strengthened, and laid down to the seabed at the offshore installation location.

a) Reel-lay method
b) J-lay method
c) Towing-lay method
d) None of the above

51. In _______, the pipe is made up at some remote location onshore, transported to the offshore installation sight by towing and laid down.

a) Reel-lay method
b) J-lay method
c) Towing-lay method
d) None of the above

52. Which of the following factors affect the installation performance of the offshore vessels?
   a) Buoyancy
   b) Stability
   c) Insubmersibility
   d) All of the above

53. The buoyancy condition for offshore vessel is
   a) Upright position
   b) Trim
   c) Heel
   d) All of the above

54. Which of the following artificial lift methods is used in deep/ultra deep water oil fields?
   a) Gas lift
   b) Electrical submersible pumping
   c) Subsea boosting
   d) All of the above

55. ________ reduces or eliminates the back pressure on the well resulting from riser hydrostatic head and the riser & flow line pressure drop caused by high viscosity. The pressure increase between output of the boosting and the backpressure on the well increase the flow from the well.
   a) Subsea boosting
   b) Gas lift
   c) Electrical submersible pumping
   d) All of the above

56. Which of the following system is the main component of subsea boosting station?
   a) Subsea gas compressor
   b) Subsea multiphase pump
   c) Subsea wet gas compressor
   d) All of the above

57. __________ platform utilizes a large diameter vertical cylinder buoy floating on the sea water surface to support the decks.
   a) Spar
   b) Barge
   c) Drillship
   d) FPSO

58. Heave causes major stability problems for ______
   a) Jack up rig
   b) Compliant towers
   c) Semi submersibles
   d) None of the above

59. Which of the following statement is true about the semi-submersible?
   a) The semisubmersible with low heave response is considered to be the most suitable
   b) The semisubmersible with high heave response is considered to be the most suitable
   c) Both (a) and (b)
   d) None of the above

60. Heave is generated in response to exposed water plane and is expressed as

a) $T = \frac{2\pi}{\sqrt{\frac{gt}{D}}}$; where T = time in seconds; t = tons per foot immersion; D= displacement in tons.

b) $T = \frac{2\pi}{\sqrt{\frac{g}{tD}}}$; where T = time in seconds; t = tons per foot immersion; D= displacement in tons.

c) $T = \frac{2\pi}{\sqrt{\frac{tD}{g}}}$; where T = time in seconds; t = tons per foot immersion; D= displacement in tons.

d) $T = 2\pi \times \sqrt{\frac{gt}{D}}$; where T = time in seconds; t = tons per foot immersion; D= displacement in tons.

61. Which of the following statement is true?

a) The lower heave response is achieved by submerging the lower hulls and floating at the column or caisson level.

b) The lower heave response is achieved by raising up the lower hulls.

c) Both (a) and (b)

d) None of the above

62. Which of the following criteria should be considered for selecting a semisubmersible?

a) Water depth and drilling depth

b) Motion characteristics and mobility

c) Environmental criteria and consumables capacity

d) All of the above

63. Which of the following method is used for the estimation of wave loads on fixed and floating offshore structure?

a) Spectral analysis method

b) Design wave method

c) Both of these

d) None of the above

64. ______ is the average wind velocity during a time interval (sampling time) of one (1) minute with a recurrence period of N years.

a) N Year Sustained Wind Velocity

b) N Year Gust Wind Velocity

c) Terminal wind velocity

d) None of the above

65. ________ is the average wind velocity during a time interval (sampling time) of three (3) seconds with a recurrence period of N years.

a) N Year Sustained Wind Velocity

b) N Year Gust Wind Velocity

c) Terminal wind velocity

d) None of the above

66. The sustained wind velocity varies with height above sea surface according to the following formula:

$V_z = V_{10}\left(\frac{z}{10}\right)^{\frac{1}{\tau}}$, Where:

a) Davenport has proposed a value of = 19 for open country, flat coastal belts, and small islands situated in large areas of water.

b) Davenport has proposed a value of = 100 for open country, flat coastal belts, and small islands situated in large areas of water.

c) Davenport has proposed a value of = 125 for open country, flat coastal belts, and small islands situated in large areas of water.

d) Davenport has proposed a value of = 7 for open country, flat coastal belts, and small islands situated in large areas of water.

67. Which of the following equation is proposed by Davenport for the determination of wind velocity?

a) $V_z = V_{10}\sqrt{0.93 + 0.007z}$

b) $V_z = V_{10}\sqrt{0.93 + 0.73z}$

c) $V_z = V_{10}\sqrt{0.93z + 0.007}$

d) $V_z = V_{10}\sqrt{0.007z + 93.0}$

68. According to the DnV Rules, the gust velocity is not to be taken less than the gust velocity calculated by following equation:

a) $(V_z)_G = V_{10}\sqrt{1.53 + 0.003z}$

b) $(V_z)_G = V_{10}\sqrt{1.53z + 0.003}$

c) $(V_z)_G = V_{10}\sqrt{5.53 + 0.003z}$

d) $(V_z)_G = V_{10}\sqrt{2.53 + 0.003z}$

69. Based on the one-seventh power law, the formula for calculating wind forces may be written by:

a) $F_w = 0.00388.C_H.C.V_{10}^2\ \ A$; where C_H is height coefficient; C is shape coefficient; V_{10} is wind velocity; A is area

b) $F_w = 0.00388.C_H.C^2.V_{10}^2\ .A$; where C_H is height coefficient; C is shape coefficient; V_{10} is wind velocity; A is area

c) $F_w = 0.388.C_H.C.V_{10}^2\ .A$; where C_H is height coefficient; C is shape coefficient; V_{10} is wind velocity; A is area

d) None of the above

70. The shape coefficient C varies from

a) 0.5 for cylindrical shapes to 1.5 for structural steel shapes

b) 1.5 for cylindrical shapes to 2.5 for structural steel shapes

c) 1.0 for cylindrical shapes to 1.5 for structural steel shapes

d) 2 for cylindrical shapes to 3.5 for structural steel shapes

71. According to DnV rules, the shape coefficient for short individual members (3 dimensional flow) is to be taken as

$$C = C_\infty\left(0.5 + 0.1\frac{L}{d}\right)$$

This formula is to be applied for:

a) $\frac{L}{d} < 0.5$

b) $\frac{L}{d} > 5.0$

c) $\frac{L}{d} < 5.0$

d) $\frac{L}{d} > 0.5$

72. The variation of current velocity with distance above bottom may, if detailed field measurements are not available, be described by the following equation:

a) $V_{cy} = V_T \cdot \left(\frac{y}{H}\right)^{\frac{1}{7}} + V_w \cdot \left(\frac{y}{H_1}\right)$

Where;

V_{cy} = Current velocity y metre above bottom, m/s;

V_T=tide-induced current in the sea surface, m/s;

V_w = wind induced velocity in the sea surface, m/s;

H_1 = Water Depth,m;

y = m above bottom;

V_{10} = sustained wind velocity, m/s

b) $V_{cy} = V_T \cdot \left(\frac{H}{y}\right)^{\frac{1}{7}} + V_w \cdot \left(\frac{y}{H_1}\right)$;

Where;

V_{cy} = Current velocity y metre above bottom, m/s;

V_T = tide-induced current in the sea surface, m/s;

V_w = wind induced velocity in the sea surface, m/s;

H_1 =Water Depth,m;

y = m above bottom;

V_{10} = sustained wind velocity, m/s

c) $V_{cy} = V_w \cdot \left(\frac{y}{H}\right)^{\frac{1}{7}} + V_T \cdot \left(\frac{y}{H_1}\right)$

Where;

V_{cy} = Current velocity y metre above bottom, m/s;

V_T = Tide-induced current in the sea surface,m/s;

V_w = wind induced velocity in the sea surface, m/s;

H_1 = Water Depth,m;

y = m above bottom;

V_{10} = sustained wind velocity, m/s

d) $V_{cy} = V_w \cdot \left(\frac{y}{H}\right)^{\frac{1}{7}} + V_T \cdot \left(\frac{y}{H_1}\right)$;

Where;

V_{cy} = Current velocity y metre above bottom,m/s;

V_T = tide-induced current in the sea surface,m/s;

V_w = wind induced velocity in the sea surface, m/s;

H_1 = Water Depth,m;

y = m above bottom;

V_{10} = sustained wind velocity, m/s

73. The following relationship exists between wave length and wave period (metric units) of a regular deep water wave:

a) $\lambda = \frac{g}{2\pi} T^2 \cong 1.56 T^2$

b) $\lambda = \frac{2\pi}{g} T^2 \cong 1.56 T^3$

c) $\lambda = \frac{g}{2\pi} T^3 \cong 1.56 T^2$

d) $\lambda = \frac{g}{2\pi} T^2 \cong 1.56 T^{1/2}$

74. The maximum height of a regular wave is given by:

a) $H_{max} = \frac{5}{7}\lambda$

b) $H_{max} = \frac{2}{7}\lambda$

c) $H_{max} = \frac{1}{7}\lambda$

d) $H_{max} = \frac{3}{7}\lambda$

75. The concept of remaining floating unanchored vessels in one spot at sea without anchors is called:
   a) Dynamic positioning
   b) Static positioning
   c) Both (a) and (b)
   d) None of the above

76. Position is defined in terms of
   a) percent of water depth
   b) height of riser
   c) volume of water displaced by floating unanchored vessel
   d) None of the above

77. Position error defines the position and stress level in the
   a) riser
   b) drill pipe
   c) Both (a) and (b)
   d) None of the above

78. The accuracy of the positioning system are about 1%, for
   a) calm water and wind
   b) Tides
   c) Both (a) and (b)
   d) None of the above

79. Maximum permissible error with respect to permissible stress level in tubular running from the sea to the ocean floor is:
   a) 5%
   b) 10%
   c) 15%
   d) 25%

80. At 5 percent water depth, the angle of vertical of the drill pipe would be
   a) 3 degree
   b) 15 degree
   c) 30 degree
   d) 60 degree

81. At 10 percent water depth, most drill string would become
   a) More stable
   b) Bent or damaged
   c) Lighter
   d) None of the above

82. The amount of error produced is closely approximated by the equation
   a) WD error = (% effective offshet) × $\frac{Heave + Tide\ motion(ft)}{[Effective\ WD(ft)]}$;

      where the % Effective Offset is the combined beacon and hydrophone offsets as a percent of water depth
   b) WD error= (% effective offshet) / $\frac{Heave + Tide\ motion(ft)}{[Effective\ WD(ft)]}$;

      where the % Effective Offset is the combined beacon and hydrophone offsets as a percent of water depth
   c) WD error= (% effective offshet) × $\frac{\text{Heave / Tide motion(ft)}}{\text{[Effective WD(ft)]}}$

where the % Effective Offset is the combined beacon and hydrophone offsets as a percent of water depth

d) WD error= (% effective offshet) × $\frac{\text{Heave} - \text{Tide motion(ft)}}{[\text{Effective WD(ft)}]}$;
where the % Effective Offset is the combined beacon and hydrophone offsets as a percent of water depth

83. For a vessel with a 20 percent effective offset in an area with a 10 foot tidal change and a 75 foot effective water depth. The tide apparent position error will be:
   a) 2.7%
   b) 5.0%
   c) 3.5%
   d) 6.8%

84. The tide or heave error is not too severe if percent offset is kept
   a) Below 20%
   b) Above 20%
   c) Below 30%
   d) Below 40%

85. Which of the following systems have the capability to provide for dynamic positioning if the vessel is within the operating range of required accuracy?
   a) The Decca
   b) Raydist
   c) Loran systems
   d) All of the above

86. _______ are used for submarine navigation.
   a) Inertial systems
   b) Non-inertial systems
   c) Combined systems
   d) None of the above

87. A marine riser operating on a floating drill rig will fail (collapse) in water depths
   a) greater than 2000 to 3000 feet if it is not partially or completely supported
   b) greater than 200 to 300 feet if it is not partially or completely supported
   c) lower than 200 to 300 feet if it is not partially or completely supported
   d) greater than 100 to 3000 feet if it is not partially or completely supported

88. Support of floating drill rig comes from
   a) Axial tension applied to the top of the riser
   b) Buoyancy along the length of the riser
   c) Both (a) and (b)
   d) None of the above

89. The axial tension for providing proper support increases
   a) With increase in water depth
   b) Riser size
   c) Mud weight
   d) All of the above

90. The rough order of magnitude (ROM) nominal tension is given by:
   a) T nominal= (weight of riser in water - weight of drilling fluid in water) x 1.20
   b) T nominal= (weight of riser in water+ weight of drilling fluid in water) x 1.20
   c) T nominal= (weight of riser in water+ weight of drilling fluid in water) / 1.20
   d) T nominal= (weight of riser in water+ weight of drilling fluid in water) x 2.50

91. Following information are given:

   Weight of 20' OD riser with ½ inch wall in water is equal to 146 lb/ft

   Drilling fluid weight = 18.0 ppg

   The nominal tension in 800 ft water depth will be:
   a) 281,280 lbs
   b) 291000 lbs
   c) 381,390 lbs
   d) 500,598 lbs

92. Which of the following type of riser is generally used during offshore drilling?
   a) Rigid TTR
   b) Rigid SCR
   c) Flexible Riser
   d) None of the above

93. Which of the following type of riser is generally used during offshore completion and workover jobs?
   a) Rigid TTR
   b) Rigid SCR
   c) Flexible Riser
   d) None of the above

94. Which of the following type of riser is generally used during offshore production operations?
   a) Rigid TTR
   b) Rigid SCR
   c) Flexible Riser
   d) All of the above

95. Which of the following type of riser is generally used during injection?
   a) Rigid TTR
   b) Rigid SCR
   c) Flexible Riser
   d) All of the above

96. Which of the following type of riser is generally used during export?
   a) Rigid TTR
   b) Rigid SCR
   c) Flexible Riser
   d) Both (b) and (c)

97. Which of the following is true about free standing hybrid riser?
   a) It consist of a vertical steel pipe tensioned by a near-surface buoyancy can
   b) A flexible jumper connects the top of the riser to the production platform
   c) Both (a) and (b)
   d) None of the above

98. Which of the following is true about the telescopic joint?
   a) It is a connection between the riser and the drilling vessel
   b) It compensates the vertical movement of drilling vessel
   c) Both (a) and (b)
   d) None of the above

99. What is the main advantage of Steel Catenary Risers?
   a) Suitable for deeper water
   b) More cost-effective alternative to flexible
   c) Both (a) and (b)
   d) None of the above

100. What is the major limitation of Steel Catenary Riser?
   a) Sensitive to waves and current loads
   b) Sensitive to vessel motion
   c) Both (a) and (b)
   d) None of the above

101. Which of the following is true about constant force tensioner system?
   a) It is attached to the top of the fixed outer barrel of the telescopic joint.
   b) It provides enough axial force in the riser string to prevent buckling
   c) Both (a) and (b)
   d) None of the above

102. Which of the following are the basic requirements for any good tensioner design?
   a) Tension capacity
   b) Wireline travel
   c) Speed response capability
   d) All of the above

103. Maximum velocity is given by:
   a) $\frac{\text{Heave (ft)} + \text{Pi(k)}}{\text{period(sec)}} = \text{Maximum velocity}\left(\frac{\text{ft}}{\text{sec}}\right)$
   b) $\frac{\text{Heave (ft)} - \text{Pi(k)}}{\text{period(sec)}} = \text{Maximum velocity}\left(\frac{\text{ft}}{\text{sec}}\right)$
   c) $\frac{\text{Heave (ft)} \times \text{Pi(k)}}{\text{period(sec)}} = \text{Maximum velocity}\left(\frac{\text{ft}}{\text{sec}}\right)$
   d) None of the above

104. Given data:
   Heave= 16 ft.; period = 16 sec.; $V_{avg.}$ = 2 ft.lsec
   What will be the maximum velocity?
   a) 3.41 ft/sec
   b) 5.5 ft/sec
   c) 9.2 ft/sec
   d) 10.2 ft/sec

105. Which of the following type of telescoping joint is used with marine risers?
   a) Constant tension system
   b) Direct axial tension system
   c) Both of these
   d) None of the above

106. __________ is located at the top of telescoping joint.
   a) Diverter
   b) Accumulator
   c) Inner barrel
   d) None of the above

107. The ______ of telescoping joint is connected to the gimbal under the drilling floor of a floating vessel.

a) inner barrel

b) outer barrel

c) Diverter

d) None of the above

108. Riser tensioners are required for water depths

a) greater than 250 ft

b) less than 250 ft

c) less than 100 ft

d) none of the above

109. The critical buckling length, L_{cr} in feet, can be calculated by:

a) $L_{cr} = 5.75\left(\frac{EI}{W}\right)^{\frac{1}{3}}$; where E is the modulus of elasticity in psi; I is the moment of inertia in in.[4]; w is the weight of the riser in water in lb.lft

b) $L_{cr} = 6.75\left(\frac{EI}{W}\right)^{\frac{1}{3}}$; where E is the modulus of elasticity in psi ;I is the moment of inertia in in.[4] ;w is the weight of the riser in water in lb.lft

c) $L_{cr} = 7.75\left(\frac{EI}{W}\right)^{\frac{1}{3}}$; where E is the modulus of elasticity in psi ;I is the moment of inertia in in.[4] ;w is the weight of the riser in water in lb.lft

d) $L_{cr} = 2.75\left(\frac{EI}{W}\right)^{\frac{1}{3}}$; where E is the modulus of elasticity in psi ;I is the moment of inertia in in.[4] ;w is the weight of the riser in water in lb.lft

110. Which of the following is a type of floating production system?

a) FPSO

b) FSO

c) TLPs

d) All of the above

111. Which of the following is the advantage of floating production system over fixed production system?

a) It reduces time from discovery to production

b) It has ability to operate in deep and ultra deep water

c) It can be used for reservoir testing in different locations

d) All of the above

112. Which of the following is the purpose of bow line?

a) Prevent backward movement

b) Keep close to pier

c) Prevent from advancing

d) Prevent from moving back

113. Which of the following is the purpose of forward breast line?

a) Prevent backward movement

b) Keep close to pier

c) Prevent from advancing

d) Prevent from moving back

114. Which of the following is the purpose of forward spring line?
   a) Prevent backward movement
   b) Keep close to pier
   c) Prevent from advancing
   d) Prevent from moving back

115. Which of the following is the purpose of after spring line?
   a) Prevent backward movement
   b) Keep close to pier
   c) Prevent from advancing
   d) Prevent from moving back

116. Which of the following is the purpose of after breast line?
   a) Prevent backward movement
   b) Keep close to pier
   c) Prevent from advancing
   d) Prevent from moving back

117. Which of the following is the purpose of stern line?
   a) Prevent backward movement
   b) Keep close to pier
   c) Prevent from moving back
   d) Prevent forward movement

118. Which of the following turret systems are best suited for harsh environment?
   a) External Turret Systems
   b) Internal Turret Systems
   c) Both (a) and (b)
   d) None of the above

119. Which of the following turret systems contain greater amount of risers?
   a) External Turret Systems
   b) Internal Turret Systems
   c) Both (a) and (b)
   d) None of the above

120. Which of the following turret system is best suited for mild to moderate environment and contain moderate number of risers?
   a) External Turret Systems
   b) Internal Turret Systems
   c) Both (a) and (b)
   d) None of the above

121. Which of the following turret system is best suited for production of reserves in the regions prone to severe conditions which might threaten the safe operation of a vessel?
   a) Buoyant Turret Mooring systems (TBM)
   b) Riser Turret Mooring systems (RTM)
   c) Both (a) and (b)
   d) None of the above

122. Spread mooring system can be used for
   a) Shallow or deepwater station keeping in mild to moderate environment
   b) Harsh and multidirectional environments where changing wid, waves and currents may impose severe loads on the anchoring system and create excessive motion on the unit

c) Both (a) and (b)
d) None of the above

123. Which of the following mooring system provides compact load and fluid transfer system?
a) Turret mooring system
b) Spread mooring system
c) Both (a) and (b)
d) None of the above

124. Spread mooring system are generally used for:
a) Ship shaped vessels
b) Semi submersibles
c) Spars
d) Both (b) and (c)

125. Which of the following statement is true about Single Buoy Mooring?
a) It is used for ship shaped vessels
b) It comprises of a round buoy fixed to the seabed by chains or anchors
c) The buoy body provides the necessary buoyancy and stability to the entire unit
d) All of the above

126. Which of the following is not a type of mooring system?
a) Single Buoy Mooring System
b) Spread Mooring System
c) Tower Mooring System
d) None of the above

127. Which of the following is a type of Single point mooring system?
a) Catenary Anchor Leg Mooring Buoy
b) Catenary Anchor Leg Rigid Arm Mooring Buoy
c) Both (a) and (b)
d) None of the above

128. Single point mooring system is most widely used because of
a) Its relatively low operating cost
b) Flexibility and reliability
c) Its ability to berth tanker of any size including the ultra large crude carriers and very large crude carriers
d) All of the above

129. Which of the following is the component of subsea production xmas tree?
a) Tree cap
b) Jumper head
c) Both (a) and (b)
d) None of the above

130. Which of the following is the basic type of the production riser?
a) J tube
b) reverse J
c) bending shoe
d) All of the above

Statement for the Linked Answer Questions 131-134

Weight of 20 inch 00 riser x 1/2inch wall on water = -33 lbs./ft. 18lbs./gal. mud = 264 lbs./ft. between ID of riser and OD of drill pipe; Nominal tension = 222 kips.

131. The safety deviser for a 4 tensioner system will be

a) 0.75

b) 0.95

c) 0.25

d) 0.15

132. The safety deviser for a 6 tensioner system will be

a) 0.75

b) 0.95

c) 0.83

d) 0.15

133. Operating tension for a 4 tensioner system will be

a) 259 kips

b) 395 kips

c) 55 kips

d) 785 kips

134. Operating tension for a 6 tensioner system will be

a) 234 kips

b) 534 kips

c) 634 kips

d) 734 kips

Answers

Q1.	c	Q28.	a	Q55.	a	Q82.	a	Q109.	d
Q2.	c	Q29.	d	Q56.	d	Q83.	a	Q110.	d
Q3.	c	Q30.	d	Q57.	a	Q84.	a	Q111.	d
Q4.	d	Q31.	d	Q58.	c	Q85.	d	Q112.	a
Q5.	a	Q32.	d	Q59.	a	Q86.	a	Q113.	b
Q6.	d	Q33.	a	Q60.	a	Q87.	b	Q114.	c
Q7.	b	Q34.	a	Q61.	a	Q88.	c	Q115.	d
Q8.	a	Q35.	c	Q62.	d	Q89.	d	Q116.	b
Q9.	c	Q36.	a	Q63.	a	Q90.	b	Q117.	d
Q10.	b	Q37.	c	Q64.	a	Q91.	a	Q118.	b
Q11.	c	Q38.	c	Q65.	b	Q92.	a	Q119.	b
Q12.	c	Q39.	a	Q66.	d	Q93.	a	Q120.	a
Q13.	a	Q40.	a	Q67.	a	Q94.	d	Q121.	c
Q14.	a	Q41.	d	Q68.	a	Q95.	d	Q122.	a
Q15.	d	Q42.	d	Q69.	a	Q96.	d	Q123.	a
Q16.	a	Q43.	b	Q70.	a	Q97.	c	Q124.	d
Q17.	b	Q44.	b	Q71.	c	Q98.	c	Q125.	d
Q18.	c	Q45.	c	Q72.	a	Q99.	c	Q126.	d
Q19.	d	Q46.	d	Q73.	a	Q100.	c	Q127.	c
Q20.	a	Q47.	a	Q74.	c	Q101.	c	Q128.	d
Q21.	a	Q48.	a	Q75.	a	Q102.	d	Q129.	c
Q22.	a	Q49.	b	Q76.	a	Q103.	a	Q130.	d
Q23.	a	Q50.	a	Q77.	c	Q104.	a	Q131.	a
Q24.	c	Q51.	c	Q78.	a	Q105.	c	Q132.	c
Q25.	a	Q52.	d	Q79.	a	Q106.	a	Q133.	a
Q26.	a	Q53.	d	Q80.	a	Q107.	a	Q134.	a
Q27.	d	Q54.	d	Q81.	b	Q108.	a		

CHAPTER 7

Petroleum Formation Evaluation

1. The process of using borehole measurement to evaluate the characteristics of sub-surface formations is called
   a) Formation evaluation
   b) Reservoir characterization
   c) Reserve estimation
   d) None of the above

2. Purpose of formation evaluation is
   a) To find a reservoir
   b) To estimate how much hydrocarbons is present
   c) To test the reservoir to see if it will produce economically
   d) All of the above

3. The practice of making a detailed record of the geologic formations penetrated by a borehole is called:
   a) Well Logging
   b) Well test analysis
   c) Reservoir simulation
   d) None of the above

4. The cost of formation evaluation may range up to
   a) 20% of total well cost
   b) 35% of total well cost
   c) 50% of total well cost
   d) 60% of total well cost

5. Which of the following analysis falls under the category of drilling operation logs (Mud log):
   a) Cutting analysis
   b) Mud analysis
   c) Drilling data collection and analysis
   d) All of the above

6. Which of the following does not fall under the category of wireline well logs?
   a) Electrical log
   b) Acoustic log
   c) Radioactive
   d) Production logs

7. Which of the following is the example of radioactive log?
   a) Induction log
   b) SP log
   c) Density log
   d) None of the above

8. Which of the following is an example of electrical log?
   a) Gamma ray log
   b) Transit time log
   c) Neutron log
   d) Spontaneous Potential (SP) log

9. Which of the following is an example of acoustic log?
   a) CBL log
   b) Neutron log
   c) Induction log
   d) SP log

10. Which of the following is an example of production log?
    a) DST log
    b) Neutron log
    c) Induction log
    d) SP log

11. Which is used to record the changes in drilling rate of penetration?
    a) Drilling time log
    b) Mud analysis
    c) Cuttings analysis
    d) None of the above

12. Which of the following indicates the presence of oil in the formation?
    a) Mud & cuttings analyses
    b) Shale factor
    c) Gas detection
    d) None of the above

13. When mud and cuttings are passed under ultra violet light fluorescence occurs. The color of fluorescence depends upon
    a) Viscosity of oil
    b) Gravity of oil
    c) Surface tension of oil
    d) None of the above

14. If the color of fluorescence is brown, the API Gravity of oil will be
    a) Less than 15
    b) Between 25 – 35
    c) Between 35 – 45
    d) Over 45

15. If the color of fluorescence is orange, the API Gravity of oil will be
    a) Between 15 – 25
    b) Between 25 – 35
    c) Between 35 – 45
    d) Over 45

16. If the color of fluorescence is yellow to cream, the API Gravity of oil will be
    a) Less than 15
    b) Between 25 – 35
    c) Between 35 – 45
    d) Over 45

17. If the color of fluorescence is white, the API Gravity of oil will be
    a) Less than 15
    b) Between 25 – 35
    c) Between 35 – 45
    d) Over 45

18. If the color of fluorescence is blue white to violet, the API Gravity of oil will be
    a) Less than 15
    b) Between 25 – 35
    c) Between 35 – 45
    d) Over 45

19. Which of the following is the component of conventional core barrel?
    a) Cutting head
    b) An outer barrel
    c) A floating inner barrel and finger type catcher
    d) All of the above

20. Which of the following is the main advantage of conventional coring?
    a) The large diameter core can be obtained for a given hole size
    b) A high percentage recovery of the formation cored
    c) No additional surface drilling equipment is required
    d) All of the above

21. Which of the following is the advantage of diamond coring?
    a) Longer bit life
    b) High percentage recovery
    c) Economic penetration of hard abrasive formation
    d) All of the above

22. Which of the following equipments are required in wireline coring in addition to usual surface drilling equipment?
    a) Wireline reel
    b) Sheave
    c) Wireline lubricator and hoisting assembly
    d) All of the above

23. Which of the following is the main advantage of wireline logging?
    a) Cutting and recovery of consecutive cores without pulling the drill pipe
    b) Alternate coring and drilling without making a trip with the drill pipe
    c) Both (a) and (b)
    d) None of the above

24. ____________ log records the electrical potential produced by interaction of formation connate water, conductive drilling fluid and certain ion selective rocks.
    a) Drilling time log
    b) Spontaneous potential log
    c) Induction log
    d) None of the above

25. In the SP curve the deflection is left if
    a) Formation water salinity is greater than mud filtrate salinity
    b) Formation water salinity is less than mud filtrate salinity
    c) Formation water salinity is equal to mud filtrate salinity
    d) None of the above

26. SP log is measured in
    a) Kilo volts
    b) Mega volts
    c) Milli volts
    d) None of the above

27. Induction logging tools can measure
   a) Formation resistivity in borehole containing oil based muds
   b) Formation resistivity in air drilled boreholes
   c) Both (a) and (b)
   d) None of the above

28. Which of the following represents Wyllie Time Average Equation?
   a) $t_{log} = \varnothing t_f + (1-\varnothing)t_{ma}$; where t_{log} = is the reading on the sonic log in μs/ft, t_{ma} is the transit time of the matrix material, t_f is the transit time of the saturating fluid
   b) $t_{log} = \varnothing t_{ma} + (1-\varnothing)t_f$; where t_{log} = is the reading on the sonic log in μs/ft, t_{ma} is the transit time of the matrix material, t_f is the transit time of the saturating fluid
   c) $t_{log} = \varnothing t_f - (1-\varnothing)t_{ma}$; where t_{log} = is the reading on the sonic log in μs/ft, t_{ma} is the transit time of the matrix material, t_f is the transit time of the saturating fluid
   d) $t_{log} = \varnothing t_{ma} + (1-\varnothing)t_f$; where t_{log} = is the reading on the sonic log in μs/ft, t_{ma} is the transit time of the matrix material, t_f is the transit time of the saturating fluid

29. Shale rocks have
   a) high resistivity
   b) low resistivity
   c) low conductivity
   d) None of the above

30. The typical formation resistivities lie between 0.5 and 1000 Ohm m²/m
   a) 0.5 and 1000 Ohm m²/m
   b) 1.5 and 1000 Ohm m²/m
   c) 5.5 and 1000 Ohm m²/m
   d) 10.5 and 1000 Ohm m²/m

31. Shaly sand formation resistivity lies between
   a) 1.5 - 50 Ohm m²/m
   b) 0.5 - 50 Ohm m²/m
   c) 3.5 - 50 Ohm m²/m
   d) 0.5 - 10 Ohm m²/m

32. Carbonates formation resistivity lies between
   a) 10 - 1000 Ohm m²/m
   b) 15 - 1000 Ohm m²/m
   c) 100 - 1000 Ohm m²/m
   d) 50 - 1000 Ohm m²/m

33. Seawater resistivity is
   a) 10.35 Ohm m²/m at 75 degrees Fahrenheit
   b) 5.35 Ohm m²/m at 75 degrees Fahrenheit
   c) 3.35 Ohm m²/m at 75 degrees Fahrenheit
   d) 0.35 Ohm m²/m at 75 degrees Fahrenheit

34. What type of log would be used to detect the presence of hydrocarbons while drilling?
   a) Mud log
   b) Sonic log
   c) Cutting analysis
   d) None of the above

35. Sonic log measures
   a) Interval transit time
   b) Velocity of sound
   c) Both of these
   d) None of the above

36. Interval transit time is dependent on
   a) Lithology
   b) Porosity
   c) Permeability
   d) Both (a) and (b)

37. The relationship of Porosity and Depth of Burial is given by:
   a) $\Phi = \Phi_0 \times e^{(-depth/\alpha)}$; where α is a compaction constant for a given region
   b) $\Phi = \Phi_0 / e^{(-depth/\alpha)}$; where α is a compaction constant for a given region
   c) $\Phi = \Phi_0 + e^{(-depth/\alpha)}$; where α is a compaction constant for a given region
   d) $\Phi = \Phi_0 - e^{(-depth/\alpha)}$; where α is a compaction constant for a given region

38. Given data: porosity at surface $(\Phi_0) = 45\%$

   Compaction constant (α) = 12,000.

   The porosity of the rock at the burial depth of 15,000 ft is:
   a) 16.0%
   b) 12.9 %
   c) 20.0 %
   d) 30.0%

39. What is the typical range of neutron porosity in a shale?
   a) 30-45%
   b) 50-60%
   c) 10-15%
   d) 5-10%

40. Which of the following is true about the Dipmeter and borehole imaging devices:
   a) Define reservoir structural and sedimentary features
   b) Find bed dip and azimuth
   c) Indicate fractures and vugs
   d) All of the above

41. The electrical logs are used to determine:
   a) fluid saturations
   b) viscosity
   c) density
   d) none of the above

42. Density in Combination with Sonic & Neutron logs gives information about:
   a) Lithology
   b) Shale Content
   c) Both (a) and (b)
   d) None of the above

43. In gas bearing formation the response of porosity by neutron log and density log will be
   a) $\Phi N < \Phi D$
   b) $\Phi N > \Phi D$
   c) $\Phi N > \Phi D$
   d) None of the above

44. Which of the following statement is true about oil signature?
    a) ΦNss>ΦDss
    b) ΦNss<ΦDss
    c) ΦNss = ΦDss
    d) None of the above

45. The Combo Tools contain:
    a) Neturon, Density, Induction
    b) Acoustic, Density, Induction
    c) Driller's time log, Density, Induction
    d) None of the above

46. The neutron, density and acoustic logs are frequently referred to as
    a) porosity logs
    b) permeability logs
    c) Both (a) and (b)
    d) None of the above

47. What type of electrical log is used to measure the resistivity at deep and medium distance from the wellbore?
    a) Induction
    b) SP log
    c) Both (a) and (b)
    d) None of the above

48. Resistivity logs are used to determine:
    a) hydrocarbon bearing zone vs water
    b) bearing zone to indicate permeable zones
    c) porosity
    d) All of the above

49. Formation bulk density (ρ_b) is a function of
    a) matrix density
    b) porosity
    c) density of the fluid in the pores
    d) All of the above

50. In a sandstone formation
    a) φN is < φD
    b) φN is = φD
    c) φN is > φD
    d) φN is > φD

51. In a limestone formation
    a) φN is < φD
    b) φN is = φD
    c) φN is > φD
    d) φN is > φD

52. In a dolomite stone formation
    a) φN is < φD
    b) φN is = φD
    c) φN is > φD
    d) φN is > φD

53. In a anhydrite formation
    a) φN is < φD
    b) φN is = φD
    c) φN is > φD
    d) φN is > φD

54. Negative density porosity is often a good indication of the presence of what?
    a) anhydrite
    b) sandstone
    c) limestone
    d) none of the above

55. Neutron logs are porosity logs that measure the ______ in a formation
   a) hydrogen concentration
   b) potassium concentration
   c) calcium concentration
   d) None of the above

56. The density log is a porosity log that measure the ________ of a formation
   a) electron density
   b) neutron density
   c) proton density
   d) None of the above

57. What general category of resistivity logs is run in salt mud?
   a) lateralogs
   b) induction logs
   c) Both (a) and (b)
   d) None of the above

58. What general category of resistivity is run in fresh mud?
   a) Induction logs
   b) SP logs
   c) lateralogs
   d) None of the above

59. Which log can be used to identify fractures near the wellbore?
   a) Microresistivity log
   b) Induction log
   c) Sonic log
   d) None of the above

60. Which type of radioactive log measures the natural radioactivity of rocks?
   a) gamma ray
   b) neutron log
   c) Density log
   d) None of the above

61. Which type of log would be used to determine hole capacity and displacement volumes?
   a) Induction log
   b) Caliper log
   c) Resistivity log
   d) Sonic log

62. Which type of log uses sound travel time to determine porosity?
   a) Resistivity log
   b) Gamma ray log
   c) Density log
   d) Sonic log

62.A lithologic log for a well is normally constructed using:
   a) drill cuttings
   b) density log
   c) gamma ray log
   d) neutron log

63. Which of the following logs is used to help identify rock porosity?
   a) Neutron log
   b) Resistivity log
   c) SP log
   d) None of the above

64. __________ measures formation density by emitting gamma radiation from the tool and records the amount of gamma radiation returning from the formation:

a) Density Log
b) Resistivity log
c) Sonic log
d) None of the above

65. The neutron log is affected the following factors:

a) hole size
b) lithology
c) hydrocarbons
d) All of the above

66. Lower apparent porosity reading is obtained from neutron log in ______

a) Oil reservoir
b) Gas reservoir
c) Both (a) and (b)
d) None of the above

67. NMR logging is used for the effective measurement of:

a) Porosity
b) Permeability
c) Both (a) and (b)
d) None of the above

68. Which log can differentiate movable water from immovable water?

a) NMR log
b) Density log
c) Porosity log
d) Gamma ray

69. __________ NMR exposes the hydrogen atoms to a second magnetic field. The energy absorbed from the second field is proportional to the density of the hydrogen nuclei, or porosity of the rock.

a) NMR Logging
b) Porosity log
c) Gamma ray log
d) Neutron log

70. Which log records electric potential between electrodes at the surface?

a) SP log
b) Neutron log
c) Gamma ray log
d) None of the above

71. SP logs mainly depends on:

a) Formation water
b) difference in salinity between drilling mud and the formation water
c) drilling mud
d) None of the above

72. __________ are used in Petrophysical & Facies Analysis

a) Radioactive Logs
b) Acoustic logs
c) Electrical logs
d) None of the above

73. __________ is used for detailed mineral identification, particularly the clay minerals.

a) SP log
b) Acoustic log

c) Natural Gamma-Ray Spectrometry

d) None of the above

74. ____________ measures the presence of three commonly occurring radioactive decay series whose parent elements are Th, U & K.

a) Neutron log

b) SP log

c) Resistivity log

d) Natural Gamma-Ray Spectrometry

75. The gamma log can be run in the

a) cased holes

b) open hole

c) Both (a) and (b)

d) None of the above

76. Laterolog is used in________

a) high-resistivity salty muds

b) low-resistivity salty muds

c) medium-resistivity salty muds

d) None of the above

77. In ________, horizontal flow of electric current into formation takes place by placing two guard electrodes above and below the current electrode and balancing the guard electrode current with that of the central generating electrode.

a) Sonic log

b) Resistivity log

c) Laterolog

d) None of the above

78. __________ is the ratio of petroleum actually formed to the genetic potential.

a) Transformation ratio

b) Formation volume factor

c) Productivity index

d) None of the above

79. If transformation ratio is greater than 0.1, it indicates

a) significant oil generation

b) significant gas generation

c) Both (a) and (b)

d) None of the above

80. Transformation ratio is determined by the

a) pyrolysis of source rock samples

b) hydrogenation of source rock samples

c) mixing of rock sample with brine solution

d) None of the above

81. Which log may help to distinguish the effects from lithology, clay content & presence of gas.

a) Acoustic log

b) Neutron log

c) Gamma ray log

d) Combined use of all the above mentioned logs

82. _________ measures the travel-time of P-wave from one end to the other end of the sonde.

a) Density log

b) Resistivity log

c) Neutron log

d) Sonic log

83. Sonic log is used in_______
   a) Open or uncased holes
   b) Cased hole
   c) Both (a) and (b)
   d) None of the above

84. Which of the following log is the only means of measuring interval velocity?
   a) Sonic log
   b) Gamma ray log
   c) Resistivity log
   d) None of the above

85. Which log is extremely useful for seismic imaging?
   a) Density log
   b) Gamma ray log
   c) SP log
   d) Sonic log

86. Which of the following statement is false:
   a) SP Log is useful when mud filtrate and formation water have similar resistivity values
   b) SP Log is useful when resistivity of mud filtrate is less than that of formation water
   c) SP Log is useful when resistivity of mud filtrate is greater than that of formation water
   d) None of the above

87. __________ uses a scintillation counter to measure the natural radioactivity of formations.
   a) Gamma ray log
   b) Caliper log
   c) Both (a) and (b)
   d) None of the above

88. Which log can be run through casing?
   a) Gamma ray log
   b) Caliper log
   c) Both (a) and (b)
   d) None of the above

89. Which of the following statement is true?
   a) gamma log reading is affected by hole diameter, so it is generally run together with a SP log.
   b) gamma log reading is affected by hole diameter, so it is generally run together with a caliper log.
   c) gamma log reading is affected by hole diameter, so it is generally run together with a sonic log.
   d) None of the above

90. The most effective log for Porosity is:
   a) NMR Logging
   b) SP log
   c) Resistivity log
   d) None of the above

91. In __________, electrical currents are sent through rocks by electrodes which record resistivity of the rocks and other fluids.
   a) Resistivity log
   b) Gamma ray log
   c) Neutron porosity log
   d) None of the above

92. The log that measures the resistivity of flushed zone R_{xo}:
   a) Microresistivity Log
   b) Laterolog
   c) SP log
   d) Gamma ray log

93. A well is going to be perforated using tubing conveyed guns run below a packer on a 3 1/2-inch tubing string. The completion interval was selected from open hole logs including induction, neutron, density, gamma ray, and spontaneous potential logs. Cased-hole log suite has been completed which included cement bond, neutron, gamma ray, and collar locator logs. The cased-hole logs are on depth with the open-hole logs. The method of correlation that would give the greatest accuracy of setting the packer and firing the guns on depth is most likely:
   a) Run a new neutron–gamma ray log to correlate with open-hole logs.
   b) Run a collar strip to correlate with the existing collar log, which is known to be on depth.
   c) Tag the plug back TD and correct based on that information.
   d) Strap-in the tubing and adjust for pipe stretch

94. A density log in a shaly sand records a formation bulk density, ρ_b = 2.20 g/cm3. The correction factor for mudcake and hole rugosity is $\Delta\rho$ = 0.051g/cm3. From measurements on cores and produced fluids it is known that the matrix and fluid densities are, ρ_{ma} = 2.68 g/cm3 and ρ_f = 1.2 g/cm3. What is the density log porosity, %?
   a) 27
   b) 29
   c) 32
   d) 36

95. Each type of logging tool has special formation evaluation applications. Which of the following statements is false?
   a) Pressure versus depth measurements can be used to identify reservoir fluids and fluid contacts
   b) NMR and density logs can be combined to predict gas-corrected formation porosity in gas-bearing zones
   c) Resistivity measurements alone can be used to differentiate oil zones from gas zones
   d) NMR fluid characterization logs can be used to identify low-resistivity pay zones.

96. A sonic logging tool recorded a travel time of 90 μs/ft in a loosely consolidated sand formation. From local experience it is known that sonic log travel times in this formation read too high because the formation is loosely consolidated. A correction factor, C_p = 1.2, should be applied to the Wyllie time-average equation to correct for lack of compaction. The matrix and formation fluid travel times are

travel time times are, Δt_{ma} = 51.0 μs/ft and Δt_f = 189 μs/ft. What is the formation porosity, %, predicted using the Wyllie equation with a compaction correction?

a) 17

b) 24

c) 28

d) 34

97. From lab measurements on core samples from a zone of interest you are given the following Archie parameters: F = 26.0, and n = 2.3. From resistivity measurements on produced formation water samples, Rw = 0.05 ohm-m at reservoir temperature. From log measurements, Rt = 10.0 ohm-m and φ = 0.20. What is Sw,,%?

a) 33

b) 35

c) 36

d) 41

98. __________ log is useful for identifying shale formations.

a) Natural gamma ray log

b) Acoustic log

c) Resistivity log

d) None of the above

99. The approximation saturation limits for dry gas production in clean sandstones are:

a) S_o = 0.2 – 0.3 ; S_g = 0.45 – 0.75; S_w = 0.30 – 0.55

b) S_o = 0; S_g = 0.45 – 0.75; S_w = 0.30 – 0.55

c) S_o = 0; S_g = 0.65 – 0.855; S_w = 0.30 – 0.55

d) S_o = 0; S_g = 0.15 – 0.75; S_w = 0.30 – 0.75

100. The approximation saturation limits for condensate production in clean sandstones are:

a) S_o = 0.01 – 0.04 ; S_g = 0.40 – 0.65; S_w = 0.30 – 0.55

b) S_o = 0; S_g = 0.45 – 0.75; S_w = 0.30 – 0.55

c) S_o = 0; S_g = 0.65 – 0.855; S_w = 0.30 – 0.55

d) S_o = 0; S_g = 0.15 – 0.75; S_w = 0.30 – 0.75

101. The approximation saturation limits for oil production in clean sandstones are:

a) S_o = 0.2 – 0.3 ; S_g = 0.45 – 0.75; S_w = 0.30 – 0.55

b) S_o = 0; S_g = 0.45 – 0.75; S_w = 0.30 – 0.55

c) S_o = 0; S_g = 0.65 – 0.855; S_w = 0.30 – 0.55

d) S_o = 0.05 – 0.30; S_g = 0.30 – 0.60; S_w = 0.30 – 0.55

102. The approximation saturation limits for water production in clean sandstones are:

a) S_o = 0.2 – 0.3 ; S_g = 0.45 – 0.75; S_w = 0.30 – 0.55

b) S_o = 0; S_g = 0.45 – 0.75; S_w = 0.30 – 0.55

c) S_o = 0; S_g = 0.65 – 0.855; S_w = 0.30 – 0.55

d) S_o = 0.0 – 0.80; S_g = 0.10 – 0.35; S_w = 0.55 – 0.90

103. Log is useful for identifying shale formations.

a. Natural gamma ray log

b. Acoustic log

c. Resistivity log

d. None of the above

104. There are known effects of gas on logging tool responses. Which of the following statements is correct?

a) Density log porosities are not affected by gas

b) Neutron log porosities are increased by gas

c) NMR log porosities are reduced by gas

d) Sonic log porosities are reduced by gas

Answers

Q1.	a	Q22.	d	Q43.	a	Q64.	a	Q85.	d
Q2.	d	Q23.	c	Q44.	b	Q65.	d	Q86.	a
Q3.	a	Q24.	b	Q45.	a	Q66.	b	Q87.	c
Q4.	a	Q25.	a	Q46.	a	Q67.	c	Q88.	c
Q5.	d	Q26.	c	Q47.	a	Q68.	a	Q89.	b
Q6.	d	Q27.	c	Q48.	d	Q69.	a	Q90.	a
Q7.	c	Q28.	a	Q49.	d	Q70.	a	Q91.	a
Q8.	d	Q29.	b	Q50.	a	Q71.	b	Q92.	a
Q9.	a	Q30.	a	Q51.	b	Q72.	a	Q93.	a
Q10.	a	Q31.	b	Q52.	c	Q73.	c	Q94.	c
Q11.	a	Q32.	c	Q53.	d	Q74.	d	Q95.	c
Q12.	a	Q33.	d	Q54.	a	Q75.	c	Q96.	b
Q13.	b	Q34.	a	Q55.	a	Q76.	b	Q97.	d
Q14.	a	Q35.	a	Q56.	a	Q77.	c	Q98.	a
Q15.	a	Q36.	d	Q57.	a	Q78.	a	Q99.	b
Q16.	b	Q37.	a	Q58.	a	Q79.	a	Q100.	a
Q17.	c	Q38.	b	Q59.	a	Q80.	a	Q101.	d
Q18.	d	Q39.	a	Q60.	a	Q81.	d	Q102.	d
Q19.	d	Q40.	d	Q61.	b	Q82.	d	Q103.	a
Q20.	d	Q41.	a	Q62.	d	Q83.	a	Q104.	c
Q21.	d	Q42.	c	Q63.	a	Q84.	a		

CHAPTER **8**

Oil and Gas Well Testing

1. The objective (s) of well testing is:
   a) To evaluate formation permeability
   b) To evaluate wellbore damage
   c) To evaluate static reservoir pressure
   d) All of the above

2. Well test analysis is a/an ________
   a) Inverse problem
   b) Forward problem
   c) Both forward and inverse problem
   d) None of the above

3. The solution to differential equations treating the well as a vertical line through a porous medium is called
   a) line source solution
   b) vertical solution
   c) constant rate terminal solution
   d) constant pressure solution

4. In ______________ the measurement and analysis of pressure data taken after an injection well is shut in.
   a) Fall off test
   b) Drill stem test
   c) Repeat formation test
   d) None of the above

5. A correctly designed well test should include
   a) Minimize test duration
   b) Have a duration exceeding the WBS effect
   c) Always be long enough to reach any suspected boundaries
   d) All of the above

6. A type of formation whose rock properties are same in all directions is called
   a) Homogeneous formation
   b) Isotropic formation
   c) Anisotropic formation
   d) None of the above

7. A type of formation with directionally dependent properties is called
   a) Isotropic formation
   b) Homogeneous formation
   c) Anisotropic formation
   d) None of the above

8. The pressure transient analysis assumes that
   a) Reservoir is homogeneous
   b) Reservoir is heterogeneous
   c) Both (a) and (b)
   d) None of the above

9. Formation with rock properties that do not change with location in the reservoir is
   a) Homogeneous formation
   b) Heterogeneous formation
   c) Isotropic formation
   d) None of the above

10. Formation with rock properties changing with location in the reservoir is
   a) Homogeneous formation
   b) Heterogeneous formation
   c) Anisotropic formation
   d) None of the above

11. Which of the following is compressible in an oil reservoir?
   a) Water
   b) Oil
   c) Pore spaces
   d) All of the above

12. The order of magnitude of compressibility for oil in an oil reservoir is
   a) 3 to 10 $\times$ 10^{-6} psi^{-1}
   b) 3 $\times$ 10^{-6} psi^{-1}
   c) 3 to 100 $\times$ 10^{-6} psi^{-1}
   d) None of the above

13. The order of magnitude of compressibility for water in an oil reservoir is
   a) 3 to 10 $\times$ 10^{-6} psi^{-1}
   b) 3 $\times$ 10^{-6} psi^{-1}
   c) 3 to 100 $\times$ 10^{-6} psi^{-1}
   d) None of the above

14. The order of magnitude of compressibility for pore spaces in an oil reservoir is
   a) 3 to 10 $\times$ 10^{-6} psi^{-1}
   b) 3 $\times$ 10^{-6} psi^{-1}
   c) 3 to 100 $\times$ 10^{-6} psi^{-1}
   d) None of the above

15. The flow equation used for vertically hydraulic fractured wells is
   a) Linear flow
   b) Spherical flow
   c) Hemispherical flow
   d) None of the above

16. The flow equation used for wells with limited perforated intervals is
   a) Linear flow
   b) Spherical flow
   c) Hemispherical flow
   d) None of the above

17. The flow equation used for a well that only partially penetrates the payzone is
   a) Linear flow
   b) Spherical flow
   c) Hemispherical flow
   d) None of the above

18. When a well is opened after shut-in condition, the production at surface is initially due to the _____
   a) Expansion of the fluid stored in the wellbore
   b) Contribution of fluid from reservoir

c) Both of these

d) None of the above

19. Which of the following method is used for analyzing drawdown test:

a) Transient method

b) Late transient method

c) Semi-steady state method

d) All of the above

20. Which of the following statement is true about steady state flow:

a) Pressure does not change with time

b) Pressure is constant with time

c) Pressure variation with time is a function of well geometry and reservoir properties

d) None of the above

21. Which of the following statement is true about semi-steady state flow:

a) Pressure does not change with time

b) Pressure is constant with time

c) Pressure variation with time is a function of well geometry and reservoir properties

d) None of the above

22. Which of the following statement is true about transient flow:

a) Pressure does not change with time

b) Pressure is constant with time

c) Pressure variation with time is a function of well geometry and reservoir properties

d) None of the above

23. The governing conditions for the diffusivity equation include:

a) The fluid is compressible

b) The formation is isotropic and homogeneous

c) The vertical permeability is equal to the horizontal permeability

d) All of the above

24. The generalized solutions to Diffusivity equation is

a) The constant-terminal-pressure solution

b) The constant-terminal-rate solution

c) Both (a) and (b)

d) None of the above

25. Which of the following parameters influence the permeability calculations?

a) Viscosity of reservoir fluid

b) Density of reservoir fluid

c) Well damage

d) None of the above

26. The skin factor can be influenced by:

a) Mud invasion

b) The perforations

c) The completion

d) All of the above

27. The objectives of Gas transient test analysis can be to evaluate:
    a) the AOF
    b) formation permeability
    c) the rate dependant skin
    d) All of the above

28. Ramey et al type curves are used for:
    a) Build up test
    b) Constant rate drawdown test
    c) Vertical fractured well with uniform flux
    d) Both (a) and (b)

29. Characteristics of constant wellbore storage effect include:
    a) Unit slope straight line on the log log plot
    b) P is a linear function of log (time)
    c) Both (a) and (b)
    d) None of the above

30. What are the characteristics of Infinite Acting Radial Flow (IARF) in a drawdown:
    a) P is a linear function of log(delta T)
    b) A horizontal line on the loglog derivative
    c) Both (a) and (b)
    d) None of the above

31. A reservoir limit test is a ________
    a) Drawdown test
    b) Build up test
    c) Injectivity test
    d) None of the above

32. A reservoir limit test is conducted to determine the ______
    a) Pore volume connected to the well
    b) Skin factor
    c) Permeability
    d) All of the above

33. The reservoir limit test would not be feasible when_______
    a) Reservoir is very large
    b) The permeability of the reservoir is very low
    c) Both of these
    d) None of the above

34. The reservoir limit test would be impractical in the ____
    a) Presence of an external drive
    b) Absence of an external drive
    c) Absence of the fault
    d) None of the above

35. What is the characteristic of Infinite Acting Radial Flow in a buildup:
    a) P is a linear function of log(-time)
    b) P is a linear function of log(tp+ delta t / delta t)
    c) We see a unit slope straight line on the loglog derivative
    d) All of the above

36. The AOF is the:
    a) Minimum gas rate of the tested well
    b) Extent of the drainage area

c) Maximum theoretical gas rate with $P_{wf} = P_{atm}$

d) Gas well productivity index

37. The reservoir limit test requires that a well that has been shut in to stabilize the reservoir pressure be produced at constant rate for a period long enough for

a) the onset of pseudo steady state of flow regime in the reservoir.

b) the transient steady state flow regime in the reservoir.

c) the steady state flow regime in the reservoir.

d) None of the above

38. Which of the example of deliverability test

a) Flow after flow test

b) Isochronal test

c) Modified Isochronal test

d) All of the above

39. Flow after flow test is best for use in____

a) High permeability reservoir

b) Low permeability reservoir

c) Both (a) and (b)

d) None of the above

40. Flow after flow test generally requires static reservoir pressure and stabilization of_____

a) 3 to 4 flow rates

b) 10 to 15 flow rates

c) 15 to 20 flow rates

d) Infinite number of flow rates

41. Isochronal test is best suited to_____

a) High permeability reservoir

b) Low permeability reservoir

c) Both (a) and (b)

d) None of the above

42. Isochronal test requires a

a) Static reservoir pressure, flow period of fixed duration followed by shut-in until pressure stabilizes again

b) Static reservoir pressure, varying flow period followed by shut-in until pressure stabilizes again

c) Static reservoir pressure, then flow and shut in periods of equal duration

d) None of the above

43. Modified Isochronal test requires a

a) Static reservoir pressure, flow period of fixed duration followed by shut-in until pressure stabilizes again

b) Static reservoir pressure, varying flow period followed by shut-in until pressure stabilizes again

c) Static reservoir pressure, then flow and shut in periods of equal duration

d) None of the above

44. Modified Isochronal test is often used for

a) Tight reservoirs

b) High volume tubing restricted and/or partially penetrated wells with fair to good permeability

c) Both (a) and (b)

d) None of the above

45. Single point test requires a ______

a) Stabilized rate and flowing pressure measured before the well is shut-in and built upto a stabilized reservoir pressure

b) Stabilized rate and flowing pressure measured after the well is shut-in and built up to a stabilized reservoir pressure

c) Stabilized rate and flowing pressure measured before the well is shut-in and do not built upto a stabilized reservoir pressure

d) None of the above

46. Single point test is widely used for

a) deliverability test where turbulence factor is known

b) deliverability test where turbulence factor is unknown

c) Both (a) and (b)

d) None of the above

47. Multi rate deliverability test

a) indicates the effect of pressure loss due to turbulence

b) doesn't indicate the effect of pressure loss due to turbulence

c) doesn't indicate the effect of pressure loss due to production of fluid

d) None of the above

48. A build up test conducted with any type of deliverability test will provide information on

a) Current reservoir pressure

b) Permeability

c) Formation flow capacity

d) All of the above

49. While designing deliverability test which of the following factors should not be considered

a) Flare efficiency

b) Time to stabilization and turbulence effect

c) Flow rates, restrictions and duration of flow

d) None of the above

50. The main objective of drawdown test is

a) To determine the average permeability of reservoir and degree of damage in the vicinity of the wellbore

b) To determine pore volume of the reservoir

c) To detect reservoir heterogeneities within the drainage area of the well

d) All of the above

51. In which region pressure transient data are dominated by well bore storage effects?

a) Early Time Region

b) Middle Time Region

c) Late transient region

d) None of the above

52. Which of the following statement is true about drawdown test?

a) Well is shut-in prior to flow test for a period of time sufficient to allow the pressure to equalize throughout the formation

b) It is not necessary to shut-in the well prior to flow test

c) Both (a) and (b)

d) None of the above

53. Which of the following statement is true about pressure build up test?

a) It describes wellbore pressure with time after well has been shut in

b) It can determine effective reservoir permeability and extent of permeability damage around the wellbore

c) It can detect presence of faults and interference between producing wells

d) All of the above

54. Which of the following assumption is usually made in pressure buildup and drawdown analyses?

a) Reservoir should be homogeneous, isotropic and horizontal of uniform thickness

b) Reservoir fluid should be slightly compressible and flow behavior should be laminar flow

c) Both (a) and (b)

d) None of the above

55. Which of the following reservoir parameter can be estimated by DST analysis?

a) Effective permeability

b) Reservoir Pressure

c) Wellbore damage

d) All of the above

56. Drill stem test is generally done in

a) Exploratory wells

b) Production wells

c) Both (a) and (b)

d) None of the above

57. Which of the following is the example of multiple well testing?

a) Interference test

b) Pulse test

c) Both (a) and (b)

d) None of the above

58. Which of the following is production testing tool?

a) Repeat formation test

b) Multiple dynamic tester

c) Both (a) and (b)

d) None of the above

59. Which of the following statement is true about Repeat Formation Test (RFT)?

a) The repeat formation test can perform a very limited transient flow and build up test at various depths

b) The RFT tool provides an estimate of formation permeability and formation damage

c) The RFT tool is capable of obtaining very accurate pressure measurements and can retrieve two fluid samples per trip in the well

d) All of the above

60. Impulse testing is useful for

a) Wells in which extended flow may not be desirable

b) Extremely tight vuggy formation where wireline formation tester fails to perform

c) Both (a) and (b)

d) None of the above

61. ____________ involves creating pressure disturbances in one well and measuring the pressure response in the other wells

a) Reservoir limit test

b) Interference test

c) Drill stem test

d) Repeat formation test

62. Which of the following is true about the importance of pressure transient test?

a) Provides estimate of near well bore condition

b) Provides estimates of distances to boundaries

c) Provides estimate of permeability under in-situ conditions

d) All of the above

63. Which of the following uses Dietz shape factor?

a) Ramey-Cobb method

b) Muskat method

c) Arps-Smith method

d) None of the above

64. Which of the following is false about radius of investigation?

a) Increasing viscosity decreases the radius of investigation

b) Increasing permeability increases the radius of investigation

c) Increasing formation volume affect the radius of investigation

d) Increasing total compressibility decreases the radius of investigation

65. In the line source solution of diffusivity equation, the exponential integral E_i is approximated by the following equation:

$$E_i(-x) = \ln(1.78x)$$

Where the argument x is given by

$$x = \frac{948 \varnothing \mu c_t r^2}{kt}$$

The above approximation is valid when:

a) Argument x is less than 0.01

b) Argument x is less than 0.1

c) Argument x is less than 1.0

d) Argument is greater than 0.01

66. Which of the following parameters influence the permeability calculations?

a) Viscosity of reservoir fluid

b) Density of reservoir fluid

c) Well damage

d) None of the above

67. An oil well is producing at a constant flow rate of 300 stb/d under unsteady state condition the reservoir has following rock and fluid properties:

B_o = 1.25 bbl/stb

$\mu_o = 1.5$ cp
$c_t = 12 \times 10^{-6}$ psi^{-1}
$K_o = 60$ md
$h = 5$ ft
$p_i = 4000$ psi
$Ø = 15\%$
$r_w = 0.25$ ft

The pressure at radii of 10 ft for 1 hour will be:

a) 3785 psi

b) 3459 psi

c) 3723 psi

d) 3922 psi

68. A well is producing at a constant flow rate of 300 STB/day under unsteady state flow condition. The reservoir has the following rock and fluid properties.

$B_o = 1.25$ bbl/stb
$\mu_o = 1.5$ cp
$c_t = 12 \times 10^{-6}$ psi^{-1}
$K_o = 60$ md
$h = 5$ ft
$p_i = 4000$ psi
$Ø = 15\%$
$r_w = 0.25$ ft

Assuming an infinite acting reservoir, what will be the bottom hole flowing pressure after one hour of production by using the dimensionless pressure approach?

a) 3759 psi

b) 3459 psi

c) 4459 psi

d) 2459 psi

Statement for linked answer questions 69-72

The following data is given for an oil well that is scheduled for a drawdown test.

Volume of fluid in the wellbore = 180 bbls

Tubing outside diameter = 2 inches

Production casing inside diameter = 7.675 inches

Average oil density in the wellbore = 45 lb/ft^3

$h = 20$ ft
$\phi = 15\%$
$r_w = 0.25$ ft
$\mu_o = 2.0$ cp
$k_o = 30$ md
$s = 0$
$c_t = 20 \times 10^{-6}$ psi^{-1}
$c_o = 10 \times 10^{-6}$ psi^{-1}

69. The wellbore storage factor caused by fluid expansion will be:

a) 0.0018 bbl/psi

b) 0.0028 bbl/psi

c) 0.0018 bbl/psi

d) 0.0058 bbl/psi

70. The wellbore storage factor caused by falling fluid level will be:

a) 0.2707 bbl/psi

b) 0.1707 bbl/psi

c) 0.3707 bbl/psi

d) 0.4707 bbl/psi

71. The total wellbore storage coefficient is:

a) 0.1825 bbl/psi

b) 0.2225 bbl/psi

c) 0.1725 bbl/psi

d) 0.1425 bbl/psi

72. The time required for wellbore storage influence to end from is:
    a) t = 115 hrs
    b) t = 225 hrs
    c) t = 325 hrs
    d) t = 55 hrs

Statement for linked answer questions 73-74

A build up test was carried out on an oil well with an estimated drainage radius of 2,460 ft. Before shut-in, the well had produced at a stabilized rate of 4900 STB/day for hours. Known reservoir data are given below:

Depth = 10,476 ft
r_w = 0.354 ft
$c_t = 22.6 \times 10^{-6}\ psi^{-1}$
B_o = 1.55 bbl/stb
μ_o = 0.20 cp
Q_o = 4900 STB/D
K_o = 60 md
H = 482 ft
p_{wf} (Δt=0) = 4000 psi
Ø = 0.09
t_p = 310 hours
casing ID = 0.523 ft
m = 40 psi/cycle
p_{1hr} = 3266 psi

73. The average permeability of the reservoir is:
    a) 12.8 md
    b) 55.5 md
    c) 70.0 md
    d) 90.0 md

74. The skin factor is:
    a) 8.6
    b) 9.2
    c) 105
    d) 2.0

Statement for linked answer questions 75-76

A well is producing oil at a rate of 800 bbl/day with a bottom hole pressure of 850 psi from a reservoir with very strong water drive (assume steady state). What is the improved permeability if a pressure drawdown analysis indicates that ΔP_{skin} at this rate is 150 psi? Other data are:

Well spacing = 40 acres, r_w = 6.5 inch, μ_o = 2.0 cP, B_o = 1.255 rb/stb; h = 45 ft; and p_e = 1350 psi.

75. The effective radius (r_e) is:
    a) 745 ft
    b) 845 ft
    c) 1045 ft
    d) 545 ft

76. The improved permeability is:
    a) 224.83 mD
    b) 124.83 mD
    c) 154.83 mD
    d) 364.83 mD

Statement for linked answer questions 77-79

In a water drive reservoir, the following data are known:

Boundary pressure = 2200 psia
Flowing pressure = 950 psia
Oil flow rate = 100 rb/day
Well radius = 0.39 ft
Effective drainage radius, r_e = 750 ft
h = 15 ft
μ_o = 0.95 cP

Flow rate (q_o) = 75 rb/day
B_o = 1.240 rb/stb

77. The Δp_{skin} is:
   a) 140.32 psi
   b) 150.52 psi
   c) 162.32 psi
   d) 175.52 psi

78. The production rate when Δp_{skin} is reduced to zero will be:
   a) 159.06 stb/day
   b) 149.02 stb/day
   c) 189.07 stb/day
   d) 199.08 stb/day

79. At r_w = 5.5 ft, value of Δp_{skin} is:
   a) 464 psi
   b) 764 psi
   c) 664 psi
   d) 864 psi

Statement for linked answer questions 80-83

Reservoir and well data are given below:

P_s = 2800 psi
p_e = 2800 psi
$C_t = 10^{-6}$ psia^{-1}
r_e = 2000 ft
r_w = 0.5 ft
μ_o = 0.65 cP
p_w = 3.050 psi
q_o = 900 stb/day
B_o = 1.150 rb/stb

80. The porosity-thickness (øh) is:
   a) 21 ft
   b) 32 ft
   c) 42 ft
   d) 51 ft

81. The permeability thickness (kh) is:
   a) 3250 mD ft
   b) 2469 mD ft
   c) 3469 mD ft
   d) 4002 mD ft

82. Assuming there is water drive (steady state flow) and bottom hole pressure remains constant. Permeability-thickness in this case will be:
   a) 3527 ft
   b) 3777 ft
   c) 3225 ft
   d) 2627 ft

83. What will be the time to reach pseudo-steady state for an oil well drilled at 40 acre spacing? Given data:

   μ = 3.75 cP, k = 45 mD, $c_{ti} = 5.5\times10^{-5}$ psi^{-1}, ø = 12.5%, A = 40 acres
   a) 15.6 days
   b) 18.5 days
   c) 20.5 days
   d) 22.5 days

Statement for linked answer questions 84-85

Following data are given for an oil well:

P_i = 1550 psi, μ = 3.75 cP, k = 45 mD, $c_{ti} = 55\times10^{-5}$ psi^{-1}, ø = 12.5%, A = 40 acres

84. The time required to reach pseudo-steady sate for an oil well drilled at 20 acre spacing is

a) 248 days
b) 322 days
c) 125 days
d) 102 days

85. The time required to reach pseudo-steady sate for an oil well drilled at 160 acre spacing is

a) 2983 days
b) 1983 days
c) 5983 days
d) 2283 days

86. The following information are given for a under saturated oil reservoir (above the bubble point):

$S_w = 17\%$,

TDS = 18 wt%,

Oil gravity = 27°API

$R_{so} = 530$ scf/stb,

Gas gravity = 0.85

$T_f = 185°F$

P = 3500 psi

$c_t = 3.6\times10^{-6}$ psi^{-1}

$T_{sep} = 75°F$

$\rho_{sep} = 115$ psia

$\rho_b = 2803$ psi

$c_o = 1.158\times10^{-5}$ psi^{-1}

$c_w = 2.277\times10^{-6}$ psi^{-1}

Assume solution gas/oil ratios do not include stock tank vent gas.

The total compressibility for the above situation will be:

a) 1.36×10^{-5} psi^{-1}
b) 2.36×10^{-5} psi^{-1}
c) 3.36×10^{-5} psi^{-1}
d) 4.36×10^{-5} psi^{-1}

87. The following information is given for a well:

$S_w = 17\%$,

$S_g = 5\%$

TDS = 18 wt%,

$R_{so} = 530$ scf/stb,

Gas gravity = 0.85

$T_f = 185°F$

P = 2000 psi

$c_f = 3.6\times10^{-6}$ psi^{-1}

$T_{sep} = 75°F$

$\rho_{sep} = 115$ psia

Oil gravity = 27°API

$c_o = 1.429\times10^{-5}$ psi^{-1}

$c_g = 5.251\times10^{-5}$ psi^{-1}

$c_w = 4.995\times10^{-5}$ psi^{-1}

Assume solution gas/oil ratios do not include stock tank vent gas.

The total compressibility for the above situation will be:

a) 1.42×10^{-4} psi^{-1}
b) 3.42×10^{-4} psi^{-1}
c) 4.42×10^{-5} psi^{-1}
d) 2.42×10^{-5} psi^{-1}

88. What will be the compressibility for low pressure, high permeability gas reservoir for the given data?

$p_b = 2803$ psi, $c_g = 8.144\times10^{-5}$ psi^{-1}, $c_w = 4\times10^{-5}$ psi^{-1}, $S_w = 0.17$, $S_g = 0.05$,

$c_f = 3.6\times10^{-6}$ psi^{-1}

a) 6.52×10^{-3} psi^{-1}

b) 6.52×10^{-5} psi^{-1}

c) 7.52×10^{-3} psi^{-1}

d) 6.52×10^{-4} psi^{-1}

89. A pressure buildup test is run on a well producing dry gas. Current average reservoir pressure is about 3,012 psia (20.77 × 103 KPa) and the flowing bottomhole pressure at the time of shut-in was 1,126 psia (7.76 × 103 KPa). Which of the following statements about analyzing the test data using a semi-log method is most accurate?

a) Horner plot of **pws** vs. Horner time ratio will be required.

b) A Horner plot of **pws2** vs. Horner time ratio will be required.

c) Bottomhole pressure must be converted to pseudopressure and then a Horner plot of pseudopressure vs. Horner time ratio will be required.

d) Semi-log analysis cannot succeed for analysis of a gas-well pressure buildup test.

90. In a low-permeability gas formation, the deliverability curve determined from a plot of **Δp2** vs. **q**, using available test data from a four-point backpressure test with equal-length flow periods, will usually not be "stabilized." Therefore, the ability of the gas well to deliver gas at a given backpressure will be:

a) About the same as predicted from the deliverability curve plotted using available test data.

b) Greater than predicted from the deliverability curve plotted using available test data.

c) Less than predicted from the deliverability curve plotted using available test data.

d) Uncorrelated to the value predicted from the deliverability plotted using available test data.

91. A producer-injector pair of wells in an infinite-acting reservoir has been producing two weeks. The distance between the wells is 500 feet (91.44 m). Assume that the injection and production rates are equal at 1,500 bpd (238.48 m3/d) and that the rock and fluid properties are the same for both injector and producer. Reservoir Properties for this formation are:

Permeability = 250 md

Porosity = .22

Net thickness = 30 ft (9.14 m)

Oil viscosity = .5 cp

Water viscosity = .5 cp

Total compressibility = 4e-5 psi (2.76e-4 KPa)

Injector Skin = 10

Producer Skin = 0

Wellbore radius of both well is 5.5 inches (139.7 mm)

Bo = Bw = 1.0 rb/stb

The expected pressure drop, psi (KPa), in the production well will be:

a) 26.91 (185.54)

b) 98.78 (681.06)

c) 239.98 (1654.60)

d) 656.42 (4525.86)

92. Given accurate test data, the exponent, n, in the Rawlins-Schellhardt gas-well deliverability equation should be

$q = C (p_r^2 - p_{wf}^2)^n$

a) Less than 0.5.

b) Between 0.5 and 1.0.

c) Greater than 1.0.

d) Between 1.0 and 2.0.

Statement for the linked answer questions 93-97

A pressure build-up test analysis for a well with q = 83 STB/D, B = 1.12 RB/STB, μ = 3.15 cp, h = 12 ft, r_w = 0.265 ft, and $p - p_{wf}$ = 265 psi gave k = 155 md and s = 2.2.

93. The pressure drop across the skin is:

a) 65 psi

b) 49 psi

c) 72 psi

d) 80 psi

94. The flow efficiency for this is well:

a) 92 percent

b) 75 percent

c) 68 percent

d) 82 percent

95. For this well the damage ratio is:

a) 1.5

b) 1.8

c) 1.22

d) 1.63

96. The damage factor of this well is:

a) 0.18

b) 0.28

c) 0.38

d) 0.48

97. The apparent wellbore radius of this well is:

a) 0.05 ft

b) 0.03 ft

c) 0.08 ft

d) 0.10 ft

Statement for the linked answer questions 98-99

The data for a well is given below:

Ø = 0.20

k = 10.0 md

μ = 20 cp

$C_t = 8.0 \times 10^{-6}\ psi^{-1}$

r_w = 0.354 ft

r_e = 1500 ft

q_o = 200 STB/D

B_o = 1.35 bbl/STB

p_i = 3000 psia

h = 100 ft

s = 0

98. The minimum producing time for the line source solution to be a reasonable approximation of the exact solution for the exponential integral solution of the diffusivity equation for a line source well in an infinite cylindrical reservoir is:

a) 0.50 hrs

b) 0.15 hrs

c) 0.35 hrs

d) 0.9 hrs

99. The time when reservoir boundaries begin to influence reservoir pressure distribution and thereby invalidate the assumption of infinite acting reservoir is:

a) 6825 hrs

b) 7500 hrs

c) 6100 hrs

d) 8100 hrs

100. In ____________, the bottom-hole pressure data is acquired after a producing well is put in shut-in condition.

a) Buildup test

b) Drawdown test

c) Drillstem test

d) None of the above

101. In ____________, the measurement and analysis of pressure data taken after a well is put on production, either initially or following an extended shut-in period.

a) Buildup test

b) Leak off test

c) Drawdown test

d) None of the above

102. Information about permeability can be obtained from______?

a) Drill stem test

b) Fracture test

c) Saturation test

d) Leak off test

103. A drawdown and buildup test has been conducted. Pressure response shows no boundaries. Semilog analyses have been conducted on both drawdown and buildup tests and a straight line slope have been identified. There is some positive skin. However, the line includes some pressures affected by wellbore storage. What effect will this have on the calculation of permeability?

a) Could be higher or lower, all depends on other factors.

b) Permeability will be underestimated in the buildup analysis, but overestimate in the drawdown analysis

c) Permeability will be overestimated in both analyses.

d) Permeability will be underestimated in both analyses.

104. $s = \left[\frac{p(r_{w}, \Delta t = 1) - p(r_w, t_p)}{m} - \log\left(\frac{k}{\emptyset c_t r_w 2} \right) + 3.23 \right]$

a) Formula to estimate skin from a buildup test

b) Continuity equation in the SI unit system

c) Well bore storage

d) Continuity equation in the field unit system

Answers

Q1.	d	Q22.	c	Q43.	c	Q64.	c	Q85.	b
Q2.	a	Q23.	d	Q44.	c	Q65.	a	Q86.	a
Q3.	a	Q24.	c	Q45.	a	Q66.	c	Q87.	a
Q4.	a	Q25.	c	Q46.	a	Q67.	a	Q88.	a
Q5.	d	Q26.	d	Q47.	a	Q68.	b	Q89.	c
Q6.	b	Q27.	d	Q48.	d	Q69.	a	Q90.	c
Q7.	c	Q28.	d	Q49.	d	Q70.	b	Q91.	b
Q8.	a	Q29.	c	Q50.	d	Q71.	c	Q92.	b
Q9.	a	Q30.	c	Q51.	a	Q72.	a	Q93.	b
Q10.	b	Q31.	a	Q52.	a	Q73.	a	Q94.	d
Q11.	d	Q32.	a	Q53.	d	Q74.	a	Q95.	c
Q12.	a	Q33.	c	Q54.	c	Q75.	a	Q96.	a
Q13.	b	Q34.	a	Q55.	d	Q76.	b	Q97.	b
Q14.	c	Q35.	b	Q56.	a	Q77.	a	Q98.	b
Q15.	a	Q36.	c	Q57.	c	Q78.	b	Q99.	a
Q16.	b	Q37.	a	Q58.	c	Q79.	a	Q100.	a
Q17.	c	Q38.	d	Q59.	d	Q80.	a	Q101.	a
Q18.	a	Q39.	a	Q60.	c	Q81.	b	Q102.	a
Q19.	d	Q40.	a	Q61.	b	Q82.	d	Q103.	d
Q20.	a	Q41.	a	Q62.	d	Q83.	a	Q104.	a
Q21.	b	Q42.	a	Q63.	a	Q84.	a		

CHAPTER 9

Health Safety and Environment in Petroleum Industries

1. EPA stands for:
   a) Environmental Protection Agency
   b) Environmental Petroleum Agency
   c) Environmental Pollution Act
   d) None of the above

2. OSHA stands for:
   a) Occupational Safety and Health Administration
   b) Overseas Safety and Health Agency
   c) Oman Safety and Health Administration
   d) None of the above

3. What is the purpose of the Safe Drinking Water Act?
   a) To regulate the type and level of contaminants in drinking water
   b) To control the pollution of surface water.
   c) Both (a) and (b)
   d) None of the above

4. What is the purpose of Clean Air Acts?
   a) To determine the EPA's responsibility to protecting and improving the nation's air quality
   b) It regulates activities that emit air pollutants
   c) Both (a) and (b)
   d) None of the above

5. What are the primary goals of the EPA?
   a) EPA stands for Environmental Protection Agency
   b) To create awareness of environmental hazards & pollutants and enforce laws in support of that
   c) Both (a) and (b)
   d) None of the above

6. What is the purpose of the RCRA (Resource Conservation and Recovery Act)?
   a) To authorize the EPA to control hazardous waste from "cradle to grave".
   b) It Authorizes EPA to seek out responsible parties for cooperation in clean up.
   c) Both (a) and (b)
   d) None of the above

7. Which of the following are the steps of an OSHA inspection:
   a) Introduction, open conference meeting

b) the inspection, closing conference

c) post-inspection review, re-inspection/conclusion

d) All of the above

8. Which of the following is the primary goal of the EPA?

a) Enforcement, which may include civil and criminal punishment

b) To create awareness of environmental hazards & pollutants and enforce laws in support of that

c) Both (a) and (b)

d) None of the above

9. OSHA has identified six basic steps of an inspection. Which step involves bringing the inspector and facility representative together to discuss the audit findings?

a) Closing conference

b) open conference meeting

c) post-inspection review

d) None of the above

10. Which of the following is true about risk identification?

a) It is the process of finding risks

b) It recognizes risks

c) It describes potential risks

d) All of the above

11. Which of the following falls under risk assessment methods?

a) Brainstorming, risk matrices, checklists

b) preliminary hazard assessment (PHA), hazard operability (HAZOP) study and failure mode and effects analysis(FMEA)

c) fault tree analysis(FTA) , management oversight and risk tree (MORT) analysis

d) All of the above

12. What are the requirements for installing and upgrading a UST?

a) The tank and piping must be installed to industry code

b) Leak detection, Spill and Overfill Protection

c) Corrosion Protection

d) All of the above

13. What is the basic requirement for leak detection?

a) Must be able to detect a leak from any portion of the tank or its piping

b) Leak detection must be installed, calibrated, operated & maintained with manufacturer's specs

c) Leak detection must meet performance requirements of federal regulations

d) All of the above

14. What is the requirement for suction piping leak detection?

a) Piping must be sloped so they drain back into storage tank when suction is released

b) Only one check valve present directly below suction pump

c) Piping must not be sloped so they can't drain back into storage tank when suction is released

d) All of the above

15. Which of the following procedure is adopted for the prevention of overfills?

a) Make sure there is enough room in the tank- stick the tank,

b) Watch entire delivery to prevent overfill/spill

c) Use overfill protection devices, Respond immediately to alarms

d) All of the above

16. Which records should be kept on UST systems?

a) Records of leak detection performance and maintenance, the required inspections & tests of corrosion protection system

b) Repaired or upgraded UST system was properly repaired/upgraded

c) Site assessment results required for a permanent closure for at least 3 years after closing a UST.

d) All of the above

17. Which of the following is a preventive measure of AST?

a) Grounding to discharge static electricity

b) Protection from weather

c) Corrosion protection isn't required

d) All of the above

18. Which of the following is true about Risk Assessment?

a) It identifies the likelihood of an adverse event

b) It identifies the hazardous events

c) It prevents the adverse events

d) None of the above

19. Which of the following steps are considered in risk Assessment?

a) Hazardous identification and dose response assessment

b) Exposure Assessment and risk Characterization

c) Both (a) and (b)

d) None of the above

20. Hazardous Identification includes:

a) Biological (bacteria) and chemical (carcinogen),

b) physical (radiation), mechanical (noise) and psychosocial (stress)

c) Both (a) and (b)

d) None of the above

21. Dose Response Assessment is:

a) Measurement of the amount of exposure and the subsequent response

b) Prediction of the likelihood adverse event

c) Both (a) and (b)

d) None of the above

22. Which of the following is true about the Exposure Assessment
    a) It identifies and describes population exposed
    b) It determines where or how the exposure occured
    c) It determines degree of exposure
    d) All of the above

23. Which of the following is true about Biomarkers?
    a) Biomarkers are chemical or genetic changes that relate to an exposure
    b) Using Biomarkers, the Volatile Organic compounds can be found in blood or breath
    c) Metals can be identified in blood, bone or hair using Biomarkers
    d) All of the above

24. Which of the following is true about Risk Characterization?
    a) It integrates information from hazard identification, dose response relationship and exposure assessment.
    b) It is performed to provide policy makers with a synopsis of all info about the nature of the risk (including uncertainties) and enable evaluation of it.
    c) It includes actions to control exposures and reduce/ avoid risk.
    d) All of the above

25. Which is the following statement is true?
    a) Risk is the probability of an adverse event occurring.
    b) Probability of an event can be expressed on a scale of 0 to 10. 0 (will not occur) and 10(certain to occur)
    c) Both (a) and (b)
    d) None of the above

26. Which of the following is true about Risk?
    a) Risk = Toxicity × Exposure
    b) Risk = Toxicity × (duration of exposure × concentration of agent)
    c) Both (a) and (b)
    d) None of the above

27. The probability of a risk occurring can be
    a) Increased by increasing the toxicity
    b) Increased by increasing duration of exposure
    c) Increased by increasing the concentration of the agent
    d) All of the above

28. Which of the following is true about Environmental Policy
    a) It is the statement of intentions by government or private organization as to its intentions toward environmental performance.
    b) Its aim is to reduce human risk and environmental damage.
    c) Both (a) and (b)
    d) None of the above

29. The main components of the Policy Cycle are
    a) Problem definition and Agenda Setting.
    b) Policy Establishment and Policy Implementation.
    c) Assessment/ evaluation of the Policy
    d) All of the above

30. Which of the following is true about National Environmental Policy Act (NEPA)
    a) Enacted in 1969,
    b) Requires federal agencies to integrate environmental values into their decision making processes.
    c) Federal agencies are required to prepare Environmental Impact Systems (EIS) to assess the environmental impact of and alternatives to major federal actions that significantly affect the environment.
    d) All of the above

31. Which of the following is true about Clean Air Act
    a) It enacted in1970
    b) It regulates air emissions from area, stationary and mobile sources.
    c) It authorizes EPA to establish National Ambient Air Quality Standards (NAAQS)
    d) All of the above

32. Which of the following statement is true about OccupationalSafety and Health Act?
    a) It enacted in 1970
    b) It ensures worker and workplace safety, under the department of labor, authorized Occupational Health and Safety Administration (OSHA) to set and enforce worker place health and safety standards.
    c) It makes recommendations for the prevention of work related indues and illness.
    d) All of the above

33. Which of the following statement is true about Toxic Substance Control Act?
    a) It enacted in 1977 and regulates discharges of water pollutants to the environment.
    b) It authorizes EPA to require an NPDES(national pollutant discharge elimination system) permit to discharge wastewater from a point source
    c) Both (a) and (b)
    d) None of the above

34. Which of the following statement is true about Toxic Substance Control Act?
    a) It enacted in1976
    b) It authorizes EPA to regulate hazardous waste from generation, transportation, treatment, storage to disposal of form its "cradle to grave".
    c) Both (a) and (b)
    d) None of the above

35. Which of the following statement is true about Resources Conservation & Recovery Act?

a) It enacted in 1980

b) It provides a superfund to clean up uncontrolled and abandoned hazardous wastes sited, accidental spills or emergency releases of pollutants.

c) It Authorizes EPA to seek out responsible parties for cooperation in clean up.

d) All of the above

36. Which of the following statement is true about Safe Drinking Water Act?

a) It is enacted in 1974

b) It deals with the protection of the quality of drinking water coming from surface or groundwater sources.

c) It authorizes EPA to set water quality standards to protect health and to regulate public water supplies to comply with standards.

d) All of the above

37. The gap analysis is

a) An assessment of the difference between requirements and the ability to meet those requirements

b) It is an assessment between a current situation and a future, desired goal

c) Both (a) and (b)

d) None of the above

38. Which of the following statement is true about health

a) It is defined as the state of complete physical, mental, and social well-being and not merely the absence of disease or infirmity.

b) It is defined as the absence of disease or infirmity only.

c) Both (a) and (b)

d) None of the above

39. Which of the following is true about toxic?

a) It is derived from the Latin word for poison,

b) the atmospheric concentration of any substance for which OSHA has established a permissible exposure limit based on a time-weighted average of eight hours may be toxic.

c) Both (a) and (b)

d) None of the above

40. Which of the following is true?

a) Safety is the mitigation of perceived or real threats within the environment.

b) A state of emotional and psychological well-being psychological health

c) Both (a) and (b)

d) None of the above

41. Which of the following is true about contamination?

a) The point contamination is the originating source of contaminates or hazards generally come from large industrial operation

b) The non-point contamination is the water pollution that originates from household activities and generate wastes that can contaminate the environment

c) Both (a) and (b)

d) None of the above

42. Which of the following statement is true

a) Brownfields are the abandoned or underutilized industrial, commercial properties with perceived or actual environmental contamination

b) Superfund site is an uncontrolled or abandoned place where hazardous waste is located, possibly affecting local ecosystems or people. A location that qualifies for the federal government's program to clean up uncontrolled hazardous waste sites.

c) Both (a) and (b)

d) None of the above

43. Which of the following is true about due diligence?

a) Doing everything possible to anticipate and avoid legal liability.

b) An independent investigation of data and/or facts relating to the transfer of the property. This inquiry is an effort to identify potential environmental contamination areas.

c) Both (a) and (b)

d) None of the above

44. Which of the following statement is true?

a) Liability is the responsibility or culpability is assigned to a party for a given action or situation.

b) Strict liability is a legal concept used to support many regulatory laws.

c) Both (a) and (b)

d) None of the above

45. The responsibility for an inspector knowing the safety regulations in any terminal belongs to?

a) The terminal staff

b) The terminal's safety director

c) The inspector

d) The inspector's dispatcher

46. Which of the following statement is true about SARA?

a) SARA stands for Superfund Amendments and Reauthorization Act;

b) It regulates reporting of storage and use of hazardous chemicals

c) Both (a) and (b)

d) None of the above

47. Which of the following is true about LTI?

a) LTI stands for lost time injury

b) It is a workplace injury that results in the worker being unable to perform his or her duties beyond the day of the injury.

c) Both (a) and (b)

d) None of the above

48. Which of the following statement is true?

a) A guide used to show the amount of outdoor air pollution and health effects is called Air quality Index

b) Points at which a substance causes problems; the beginning of symptoms showing that there is a problem is known as Thresholds

c) A distribution model is the assumption that each individual has a threshold and tolerance level different from another individual regarding risk of harm from environmental hazards

d) All of the above

49. Which of the following agencies devised a tracking system for data collection on environmental issues?

a) Environmental Protection Agency and Centers for Disease Control

b) Division of Environmental Hazards and Health Effects

c) Department of Health and Human Services

d) All of the above

50. Which of the following statement is true?

a) In occupational health, stress is defined as a heightened physical and emotional reaction to an environment, event, person, or memory that is uncomfortable or unpleasant

b) Biological hazards occur when pathogens and other living organisms that cause acute and chronic infections. The most common way pathogens enter the body is through a break in the skin or through mucous membrane passageways

c) Both (a) and (b)

d) None of the above

51. Which of the following is not one of the three phases of an accident, according to the Haddon model?

a) Preliminary phase

b) Pre-injury phase

c) Injury phase

d) Post-injury phase

52. Which of the following types of prevention would be in effect when personnel participate in "mock situations" or "trial runs"?

a) Primary prevention

b) Secondary prevention

c) Tertiary prevention

d) Risk management strategies

53. Which of the following is not a risk assessment model?

a) Environmental hazard model

b) Epidemiologic model

c) Exposure-disease model

d) None of the above

54. If you need to know specific safety or exposure information about benzene you should consult?

a) Your doctor or pharmacist

b) Your Safety Manual

c) Material Safety Data Sheet

d) DOT Emergency Response Guide

55. What does an explosion meter measure?

a) The amount of oxygen in a space

b) Whether a space is safe for entry

c) Whether or not there is an explosive mixture in a space, capable of supporting combustion

d) The flash point of a gas mixture

56. P.P.E. stands for

a) Private Pension Estimate

b) Personal Protective Equipment

c) Private Protection Equipment

d) It has no meaning in terms of safety

57. Which one of the following list contains only PPE?

a) Dusk mask, safety goggles and ear defenders

b) Ear defenders, reading glasses and trainers

c) Thermal gloves, machine guard and overalls

d) Machine guard, dusk mask and safety goggles

58. The minimum personal protective equipment required when sampling is?

a) Gloves, respirator, hard hat and SCBA

b) Gloves, goggles, hard-hat, uniform & work shoes

c) Gloves, face-shield and sun glasses

d) Gloves, uniform and SCBA

59. What is the first action to take when approaching the scene of an accident?

a) Checks for signs of life

b) Wait for the emergency services

c) Attend to most serious injury

d) Make sure the area is safe to enter

60. What is the best way of dealing with a hazard to ensure others are not put at risk?

a) Remove it immediately

b) Leave it for others to sort out

c) Place a barrier tape around it

d) Display a notice or warning sign

61. There is a fault with the PPE, the user should?

a) Not do the task and inform the supervisor of the problem immediately

b) Get somebody else to complete the task

c) Use the equipment and continue the task anyway

d) Use the equipment to do the task and tell the supervisor later

62. The most important reason for wearing the correct type of gloves while sampling is so that–

a) The sample is not contaminated

b) Your hands do not get dirty

c) Hazardous chemicals are not absorbed through your skin

d) None of the above

63. When lifting anything heavy, which muscles should take most of the weight?

a) Upper arm muscles

b) Leg muscles

c) Back muscles

d) All of the above

64. Which of these best describes a hazard?

a) Something that is potentially harmful

b) Tools lying around on the floor

c) Cables lying around on the floor

d) Unlabelled chemicals

65. The most informative source of information about the hazards of a product or chemical is –

a) The job sheet

b) The Bill of Lading

c) The (Material) Safety Data Sheet

d) The dispatcher

66. What do the initials "(M) SDS" stand for?

a) (Material) Storage and Distribution System

b) (Material) Safety and Distribution Sheet

c) (Material) Storage and Data System

d) (Material) Safety Data Sheet

67. On which of the following documents would you find a "CAS" Number?

a) A (Material) Safety Data Sheet, (M)SDS

b) A Bill of Lading

c) A Certificate of Analysis

d) A Chemical compatibility list

68. In which year was the Health and Safety at work act enacted?

a) 1947

b) 1974

c) 1957

d) 1975

69. Wet chemical fire extinguishers are identified by which colour?

a) Red

b) Black

c) Cream

d) Yellow

70. When gauging a tank that is emitting vapors, you should position yourself?

a) It is not important

b) Up-wind of the gauge hatch

c) With the wind in your face

d) With the wind at your left or right side

71. What is a "CAS Number"?
   a) A communication and shipping number assigned to a product by the manufacturer
   b) A chemical formula number used to identify the product
   c) A unique identifying number assigned to a product by the Chemical Abstract Service
   d) An identifying number used by manufacturers to assign categories of chemicals for sale

72. Which of the following is true about the flaring?
   a) It is controlled burning of natural gas that cannot be processed for sale or use because of technical or economical reasons
   b) It is done during emergencies like power failures, equipment failures etc. to prevent any accident
   c) Both (a) and (b)
   d) None of the above

73. Which of the following gas contribute to global warming?
   a) CO_2
   b) CH_4
   c) Both (a) and (b)
   d) None of the above

74. Which of the following statement is true?
   a) Pretreatment of effluent water may be done in two corrugated plate inceptors operating in parallel to accept the increased flow rate
   b) It can handle the separation of free oil and suspended solids from water
   c) The inceptors reduce the gross separable oil contamination to a level acceptable for feeding to the pressure sand filter
   d) All of the above

75. The primary treatment of the effluent water is usually done by
   a) Sand filtration
   b) Biofiltration
   c) Clarifier
   d) All of the above

76. The sand filter filters
   a) suspended solids from the effluent water
   b) suspended oil from the effluent water
   c) Both (a) and (b)
   d) None of the above

77. Which of the following is true about secondary treatment of waste water?
   a) Secondary treatment of the filtered effluent water is done using Biotowers filled with randomly packed cylindrical type hollow plastic media (polypropylene) as substrate for biomass (biological filtration)
   b) Biotowers are designed to reduce BOD of the effluent water

c) Nutrient solution (containing N_2 and phosphorous) are added up stream of biotowers to maintain the healthy biomass growth

d) All of the above

78. Corrugated plate inceptors are used

a) For the treatment of solid wastes

b) For the treatment of air pollutants

c) For the treatment of waste water

d) None of the above

79. Which of the following is true about tertiary treatment of waste water?

a) Tertiary treatment of biotower effluent is done in clarifiers

b) It removes the biological solids

c) It ensures the MINAS suspended solid limit.

d) All of the above

80. Which of the following is true about the final treatment of waste water?

a) Final treatment of the effluent water is done in guard pond by supplying low pressure compressed air from shore mounted air blower.

b) Suspended bacteria oxidizes any residual BOD present in effluent water in presence of oxygen, thus maintaining the BOD level to the required value of MINAS.

c) Both (a) and (b)

d) None of the above

81. The pH requirement of effluent water for disposal (as per MINAS) is

a) 6.0 – 8.5

b) 9.0 – 11 .5

c) 3.0 – 7.5

d) 6.5 – 8.0

82. The maximum TSS requirement of effluent water for disposal (as per MINAS) is

a) 50 ppm

b) 20 ppm

c) 45 ppm

d) 15 ppm

83. The BOD of effluent water for disposal (as per MINAS) should be

a) 35 ppm

b) 15 ppm

c) 45 ppm

d) 55 ppm

84. The COD of effluent water for disposal (as per MINAS) should be

a) 250 ppm

b) 350 ppm

c) 150 ppm

d) 500 ppm

85. The concentration of oil & grease in effluent water for disposal (as per MINAS) should be

a) 10 ppm

b) 30 ppm

c) 100 ppm

d) 200 ppm

86. The concentration of sulphides in effluent water for disposal (as per MINAS) should be
    a) 1.5 ppm
    b) 0.5 ppm
    c) 50 ppm
    d) 100 ppm

87. The minimum dissolved oxygen concentration in effluent water for disposal (as per MINAS) should be
    a) 50 ppm
    b) 100 ppm
    c) 5 ppm
    d) 90 ppm

88. All portable electronic equipment must be ________ before use?
    a) Checked
    b) Cleaned
    c) Calibrated
    d) Grounded

89. The best source of information about the hazards of any product being inspected is?
    a) The inspector's dispatcher
    b) A knowledgeable chemist
    c) The inspector's previous experience
    d) (Material) Safety Data Sheets

90. Which of the following is true about the flaring?
    a) It is controlled burning of natural gas that cannot be processed for sale or use because of technical or economical reasons
    b) It is done during emergencies like power failures, equipment failures etc. to prevent any accident
    c) Both (a) and (b)
    d) None of the above

91. Which of the following is true?
    a) LEL stands for Lower Explosive Limit
    b) LEL is the minimum proportion of hydrocarbons in the hydrocarbon air mixture such that an explosion or fire could occur in the presence of a source of ignition such as spark or an open flame.
    c) It is given as % vol/vol air
    d) All of the above

92. An explosimeter measuring LEL% is utilized to sample the atmosphere within a cargo tank and a reading of 15% is observed. What does the reading mean?
    a) The atmosphere in the tank contains 15% oxygen
    b) The atmosphere in the tank is 15% towards an explosive mixture of air and hydrocarbon vapors
    c) The atmosphere in the tank is 15% hydrocarbon vapors
    d) The atmosphere in the tank is a 15% mixture of air and hydrocarbon vapors

93. Which of the following is true?
    a) HEL stands for high explosive limits

b) HEL is the maximum proportion of hydrocarbons in the hydrocarbon air mixture such that an explosion or fire could occur in the presence of a source of ignition such as spark or an open flame.

c) It is given as %vol/vol air

d) All of the above

94. What does an oxygen meter measure?

a) The percentage of oxygen below the LEL of a hydrocarbon/air mixture

b) The amount of oxygen needed to make a confined space safe for entry

c) The percentage of oxygen contained in the atmosphere being sampled

d) None of the above

95. The LEL% vol/vol for methane is

a) 5.3

b) 8.5

c) 15.0

d) 20.5

96. The HEL%vol/vol for methane is

a) 14

b) 25

c) 52

d) 31

97. The sense of smell is not reliable for detecting hydrogen sulfide because–

a) It is difficult to detect by sense of smell

b) The level at which you can smell it is above the permissible exposure limit

c) At 100 ppm a person's sense of smell is deadened within minutes, thereby giving that person a false sense of security

d) You might have a cold and be unable to breath through your nose

98. Which of the following conditions can cause false readings on an explosion meter?

a) Low hydrocarbon vapor content

b) High moisture content

c) Low oxygen content

d) Both b and c

99. Which of the following exposure limit for H_2S [for use with respirators] are correct?

a) 500 ppm maximum exposure when using a full face respirator

b) 100 ppm maximum exposure when using a half-mask respirator

c) A self contained breathing apparatus must be used if the exposure exceeds 500 ppm

d) Only a self-contained breathing apparatus is permissible for any exposure above the permissible exposure limit

100. ______solution is added in the upstream of Boitowers as a nutriants for bacteria.

a) Diammonium phosphate

b) Hydrochloric acid

c) Sulphuric acid

d) None of the above

101. When working in an environment containing over 10ppm of H_2S the only effective protection is?

a) Full face respirator

b) Half mask respirator

c) Self-contained breathing apparatus (SCBA)

d) All of the above

102. Water spray system is provided for

a) LPG spheres

b) Crude oil storage tanks

c) Both (a) and (b)

d) None of the above

103. Before taking a gauge, static electricity can be discharged from your body by?

a) Using a tank gauge meter

b) Touching a grounded structure such as a tank railing, with bare hands

c) Use of natural fiber sampling cords

d) Touching a grounded structure such as a tank railing, while wearing rubber gloves

104. Each crude oil storage tanks should be equipped with

a) Fire water sprinklers

b) Foam Pourer Lines

c) Both (a) and (b)

d) None of the above

105. Which of the following is Ancillary Fire Fighting Equipments/ Consumables?

a) Fire fighting hose

b) Asbestos Fire Suit

c) Breathing apparatus with O_2 cylinder

d) All of the above

106. Which of the following is true?

a) For butane gas emission, the sensor of Gas Detection System should be located above the probable leak source.

b) For butane gas emission the sensor of Gas Detection System should be located below the probable leak source.

c) For butane gas emission, the sensor of Gas Detection System should be located anywhere.

d) None of the above

107. Which of the following is true?

a) For methane gas emission, the sensor of Gas Detection System should be located above the probable leak source.

b) For methane gas emission the sensor of Gas Detection System should be located below the probable leak source.

c) For methane gas emission, the sensor of Gas Detection System should be located anywhere.

d) None of the above

108. What do the initials "UEL" stand for?

a) Upper environmental level
b) Unknown environmental level
c) Upper explosive limit
d) Unknown explosive levels

109. Hydrogen sulfide can be recognized by the following characteristics?

a) Pale yellow gas with a sweetish taste and strong pungent odor
b) Pale yellow gas with the unpleasant odor of "Rotten Eggs"
c) Colorless gas with a sweetish taste and the unpleasant odor of "Rotten Eggs"
d) Colorless gas with little or no odor

110. The permissible exposure level of hydrogen sulfide is?

a) 10 ppm
b) 15 ppm
c) 25 ppm
d) 50 ppm

111. The principal limitation of a filter or cartridge respirator is?

a) It does not supply oxygen
b) The face piece tends to fog up
c) The expense to replace the cartridges
d) Keeping the face piece from sweating up

112. The permissible exposure level of H_2S is

a) 10 ppm
b) 1 ppm
c) 5 ppm
d) ppm

113. If the atmosphere in a cargo tank is stated to be "below the LEL" what does this mean?

a) There is too much oxygen in the tank to support combustion
b) There are not enough hydrocarbon vapors in the tank to permit combustion
c) Neither of the above
d) Both a and b are correct

114. The appearance and characteristic odor of benzene is_____

a) Clear colorless liquid with a sweet odor
b) Clear colorless liquid with a sour odor
c) Light brown liquid with no distinguishable odor
d) Light brown liquid with a strong pungent odor

115. How can benzene enter your body?

a) By inhalation
b) By absorption through the skin
c) By Ingestion
d) All of the above

116. The minimum required respirator for working in a benzene environment that is greater than the permitted exposure limit is?

a) Self contained breathing apparatus

b) Full face respirator with an organic vapor cartridge

c) Half-mask respirator with an organic vapor cartridge

d) Half-mask respirator with an acid gas cartridge

117. When working with benzene, in addition to a respirator, what other personnel protective equipment is required?

a) Safety glasses, hard hat and leather gloves

b) Long-sleeve coveralls and leather boots

c) Rubber gloves, goggles, chemical protection suit and hard hat

d) Personal Protective Equipment is not required

118. Which of the following maximum benzene concentration limit is correct, when using a respirator?

a) 50 ppm maximum exposure when using a full-face respirator

b) 10 ppm maximum exposure when using a half-mask respirator

c) A self contained breathing apparatus must be used if the exposure exceeds 50 ppm

d) All of the above

119. The permissible exposure limit for benzene is an 8-hour time weighted average of?

a) 10 ppm

b) 1 ppm

c) 5 ppm

d) 0.5 ppm

120. Workers who are exposed to benzene at or above the action level for at least 30 days per year must have a medical examination?

a) Every six months

b) Only if they exhibit signs of exposure

c) Only if the exposure exceeds ten times the permissible exposure limit

d) Once a year

121. The effects of acute exposure [short term] to benzene are?

a) Shortness of breath, irritability, headache, nausea, dizziness, intoxication

b) Irritation of the eyes, nose and respiratory tract

c) Convulsions and loss of consciousness

d) All of the above

122. Which of the following statement is true?

a) Primary treatment of waste water is done for bulk oil removal

b) Secondary treatment is biological treatment

c) Tertiary treatment is done for TSS and COD removal

d) All of the above

123. Which of the following is potentially hazardous component of drilling fluids?
   a) Oil
   b) Salts
   c) Soluble trace elements consisting of Zn, Pb, Cu, Cd, Ni, Hg
   d) All of the above

124. Which of the following are the disposal methods for spent drilling fluids?
   a) solid encapsulation,
   b) pumping into safe formations
   c) liquid extraction and chemical treatment
   d) All of the above

125. Water based fluids may be spread directly over adjacent agricultural or forest areas after
   a) The adjustment of pH
   b) The adjustment of ion content
   c) Both (a) and (b)
   d) None of the above

126. Which of the following are the solid removal equipments for spent drilling fluids
   a) shale shaker
   b) desander
   c) desilter & centrifuge
   d) All of the above

127. Which of the following are waste of concerned?
   a) drilling fluid and drill cuttings
   b) produced water
   c) radioactive materials
   d) All of the above

128. Which of the following are the cleanup methods for oil spill?
   a) Biodegradation
   b) controlled burning
   c) Dispersant and skimming
   d) All of the above

129. Which of the following are the sources of hazardous noise
   a. Mud pumps and tanks, Shale shaker/centrifuge
   b. Offloading, main oil and cement pumps
   c. Fracturing
   d. All of the above

130. Which of the following is true about hearing loss?
   a) Noise induced hearing loss is the permanent hearing loss caused by long term exposure to hazardous noise
   b) The severity of hearing loss is affected by the intensity of noise and duration of exposure
   c) Both (a) and (b)
   d) None of the above

131. __________ is an enclosed or partially enclosed area that is big enough for a worker to enter.
   a) Confined space
   b) Open space
   c) Smaller space
   d) None of the above

132. A confined space is one that?

a) Has limited means of access and exit

b) Is not designed for continuous occupation

c) Has limited natural ventilation

d) All of the above

133. Which of the following is considered to be confined spaces?

a) A cofferdam

b) An external floating roof tank

c) A ship's pump room

d) All of the above

134. Confined spaces are common in oil and gas industry, particularly in processing operations. Confined spaces include

a) storage tanks, process and reaction vessels,

b) boilers, ventilation and exhaust ducts,

c) tunnels and pits and pipelines

d) All of the above

135. Before entering confined spaces which of the following tests is required?

a) Oxygen content

b) Lower explosive limit

c) Toxic vapor testing

d) All of the above

136. Which of the following is an example of confined space?

a) Cargo tank

b) Grain silo

c) Pump room on a ship

d) All of the above

137. What type of respirator filter cartridges should you use for protection against hydrogen sulfide?

a) Acid Gas

b) None. Only a self contained breathing apparatus is acceptable

c) Organic Vapor

d) Radionuclides, highly toxic dusts, mists and fumes

138. NORM can be found in many components of operating oil and gas facilities including

a) piping runs, including down hole piping,

b) well heads, production manifolds,

c) gas/oil separator flow lines, dehydrator and desalinator, valves, storage tanks

d) All of the above

139. A chemical has a strong odor, this is?

a) An indication that a hazard exists

b) Indicates low vapor concentration

c) Indicates high vapor concentration

d) An unreliable source of specific information regarding the chemical

140. Work activities that may carry a risk of exposure to mercury in gas processing facilities include

a) vessel cleaning, welding and grinding,

b) buffing and polishing,

c) machining and pipefitting

d) All of the above

141. Which of the following is defined as a corrosive liquid?

a) An acid solution

b) A caustic solution

c) Neither of the above

d) Both a and b

142. Workers performing the following activities are at risk of breathing silica dust by

a) abrasive blasting using silica containing products,

b) cementing operations,

c) drilling using dry product additives that contain quartz, hydraulic fracturing

d) All of the above

143. Products have defined limits of combustion. These are?

a) The Permissible Exposure Limit and the Threshold Limit Value

b) The Lower Explosive Limit, Upper Explosive Limit and the Flash Point

c) The Flash Point and the Threshold Limit Value

d) The Permissible Exposure Limit and the Lower Explosive Limit

144. What is the safe oxygen content range in a confined space?

a) Between 19.5% and 23.5%

b) Between 18.6% and 20%

c) Between 19% and 25%

d) None of the above

145. The opportunity for a build up of a static electricity charge can be reduced by?

a) Wearing rubber gloves

b) Not allowing your hands to slide on the hand rail

c) Grounding yourself and your sampling equipment before opening the gauge/ sample hatch

d) Use of stainless steel equipment

146. To help prevent a build up of static electricity?

a) Always use a sampling cord made of synthetic fiber

b) Always tie the end of the sampling cord to the railing of the tank

c) Use a sampling cord that contains no synthetic fiber

d) Hold the sample cord against the gauge hatch while pulling up the sample

147. To avoid the buildup of static electricity when using a portable electronic thermoprobe (PET)?

a) Hold on to the railing or other metal part of the tank while using the PET

b) Attach the ground wire of the PET to the tank before

opening the gauge hatch then slowly lower the probe assembly into the oil

c) Since the probe is plastic and does not conduct electricity, no static electricity can form

d) Any of the above

148. Why should pumping be suspended when first-in or one-foot samples are taken?

a) To allow any gas to dissipate

b) To allow static electricity to dissipate

c) So the vapors don't blow in your face

d) To give time for analysis results

149. Some silica containing materials include

a) concrete

b) asphalt

c) hydraulic fracturing sand

d) All of the above

150. The most hazardous component of sour crude oil is?

a) The smell

b) Low flash point

c) Toxicity

d) Hydrogen sulfide

Answers

Q1.	a	Q31.	d	Q61.	a	Q91.	d	Q121.	d
Q2.	a	Q32.	d	Q62.	c	Q92.	b	Q122.	d
Q3.	c	Q33.	c	Q63.	b	Q93.	d	Q123.	d
Q4.	a	Q34.	c	Q64.	a	Q94.	c	Q124.	d
Q5.	c	Q35.	d	Q65.	c	Q95.	a	Q125.	c
Q6.	c	Q36.	d	Q66.	d	Q96.	a	Q126.	d
Q7.	d	Q37.	c	Q67.	a	Q97.	c	Q127.	d
Q8.	c	Q38.	a	Q68.	b	Q98.	d	Q128.	d
Q9.	a	Q39.	c	Q69.	d	Q99.	d	Q129.	d
Q10.	d	Q40.	c	Q70.	d	Q100.	a	Q130.	c
Q11.	d	Q41.	c	Q71.	c	Q101.	c	Q131.	a
Q12.	d	Q42.	c	Q72.	c	Q102.	c	Q132.	d
Q13.	d	Q43.	c	Q73.	c	Q103.	b	Q133.	d
Q14.	d	Q44.	c	Q74.	d	Q104.	c	Q134.	d
Q15.	d	Q45.	c	Q75.	a	Q105.	d	Q135.	d
Q16.	d	Q46.	c	Q76.	c	Q106.	b	Q136.	d
Q17.	d	Q47.	c	Q77.	d	Q107.	a	Q137.	b
Q18.	a	Q48.	d	Q78.	c	Q108.	c	Q138.	d
Q19.	c	Q49.	d	Q79.	d	Q109.	c	Q139.	d
Q20.	c	Q50.	c	Q80.	c	Q110.	a	Q140.	d
Q21.	a	Q51.	a	Q81.	a	Q111.	a	Q141.	d
Q22.	d	Q52.	b	Q82.	b	Q112.	a	Q142.	d
Q23.	d	Q53.	a	Q83.	b	Q113.	b	Q143.	b
Q24.	d	Q54.	c	Q84.	a	Q114.	a	Q144.	a
Q25.	c	Q55.	c	Q85.	a	Q115.	d	Q145.	c
Q26.	c	Q56.	b	Q86.	b	Q116.	c	Q146.	c
Q27.	d	Q57.	a	Q87.	c	Q117.	c	Q147.	b
Q28.	c	Q58.	b	Q88.	d	Q118.	d	Q148.	b
Q29.	d	Q59.	d	Q89.	d	Q119.	a	Q149.	d
Q30.	d	Q60.	d	Q90.	c	Q120.	d	Q150.	d

CHAPTER **10**

Enhanced Oil Recovery Techniques

1. Which of the following process refers to the recovery of oil through the injection of fluids and energy not normally present in the reservoir?
   a) EOR
   b) Artificial Lift Techniques
   c) Work over
   d) None of the above

2. Which of the following statement is true?
   a) The ultimate aim of EOR processes is to increase overall oil displacement efficiency
   b) Oil displacement efficiency is the function of microscopic and macroscopic displacement efficiency
   c) Microscopic efficiency can be increase by reducing interfacial surface tension between displacing fluid and oil
   d) All of the above

3. The oil produced by the reservoir drive energy is called
   a) primary production
   b) improved oil recovery
   c) ultimate oil recovery
   d) None of the above

4. Most widely used post-primary recovery is
   a) Water flooding
   b) Polymer flooding
   c) Steam flooding
   d) None of the above

5. What is the depth limit for steam flooding?
   a) 3000 ft
   b) 5000 ft
   c) 6000 ft
   d) 10000 ft

6. Steam flooding should be used for
   a) pay zones greater than 20 ft thick
   b) pay zones less than 20 ft thick
   c) both (a) and (b)
   d) None of the above

7. Reservoir volumes of oil mobilized by EOR agent divided by reservoir volumes contacted by EOR agent is called
   a) displacement efficiency
   b) volumetric sweep efficiency
   c) vertical sweep efficiency
   d) None of the above

8. Reservoir volumes of oil contacted by displacing agent divided by reservoir volumes of oil initially in place is called
   a) displacement efficiency
   b) volumetric sweep efficiency
   c) vertical sweep efficiency
   d) None of the above

9. What depends on heterogeneity, gravity, mobility ratio, capillary forces, and injected volumes?
   a) Displacement efficiency
   b) Vertical sweep efficiency
   c) Volumetric sweep efficiency
   d) None of the above

10. The area contacted by displacing agent divided by the total area is called
   a) area sweep efficiency
   b) Volumetric sweep efficiency
   c) Both (a) and (b)
   d) None of the above

11. The limitations of the CO_2 flooding include
   a) Low viscosity
   b) Availability
   c) Surface facilities
   d) All of the above

12. Which of the following method develops miscibility to displace oil from reservoir
   a) CO_2 flooding
   b) Polymer flooding
   c) Water flooding
   d) None of the above

13. The method consists of injecting large quantities of gas that may be miscible or immiscible depending on the pressure and oil composition is
   a) Nitrogen/flue gas flooding
   b) Water flooding
   c) Polymer flooding
   d) All of the above

14. The method consists of adding water soluble polymers to the water before it is injected into the reservoir is called
   a) Water flooding
   b) Polymer flooding
   c) Steam flooding
   d) None of the above

15. The surfactant/polymer flooding doesn't improve:
   a) displacement efficiency
   b) volumetric sweep efficiency
   c) both (a) and (b)
   d) None of the above

16. The method consists of injecting steam to heat the reservoir, thus reducing the oil viscosity and displace the oil is known as
   a) Steam flooding
   b) CO_2 flooding
   c) N_2 flooding
   d) None of the above

17. Which of the following is false in case of water flooding?
   a) It improves areal sweep efficiency
   b) It decreases pressure or pressure maintenance

c) It increases production rate

d) All of the above

18. How do you minimize trapped oil saturation?

a) increase capillary number

b) decrease capillary number

c) increase viscosity of the oil

d) None of the above

19. What are the types of heterogeneity?

a) areal variations

b) vertical variations

c) fractures

d) All o f these

20. The method that describes layers according to their relative position within the vertical rock column is known as

a) Position method

b) Layer method

c) Vertical Column method

d) None of the above

21. __________represents the relative effect of viscous forces and surface tension acting across the interface between two immiscible liquids.

a) capillary number

b) Reynolds number

c) Formation volume factor

d) None of the above

22. What can mitigate the effect of heterogeneity and gravity?

a) Capillary number

b) favorable mobility ratio

c) both (a) and (b)

d) None of the above

23. Which of the following statement is true?

a) Mobility ratio greater than 1 is favorable for displacement

b) Mobility ratio less than 1 is favorable for displacement

c) Mobility ratio greater than 2 is favorable for displacement

d) Mobility ratio is equal to 0 is favorable for displacement

24. Gravity forces are strong compared to viscous forces when the

a) Capillary number (Nc) is large.

b) Capillary number is smaller

c) Both (a) and (b)

d) None of the above

25. Application of engineering techniques to produce more oil after primary production is known as

a) improved oil recovery

b) ultimate oil recovery

c) Primary recovery

d) None of the above

26. The total production from a well or field primary production and improved oil recovery that is justified by economics is known as

a) improved oil recovery

b) ultimate oil recovery

c) Primary recovery

d) None of the above

27. The well pattern having 4 injectors and 1 producer is called
    a) five spot
    b) line drive
    c) Seven spot
    d) Four spot

28. The well pattern having 6 injectors and 1 producer is called
    a) five spot
    b) line drive
    c) Seven spot
    d) Four spot

29. The well pattern having 6 injectors and 1 producer is called
    a) Seven spot
    b) line drive
    c) Inverted seven spot
    d) Four spot

30. The well pattern having 2 injectors and 2 producers is called
    a) Direct line drive
    b) Four spot
    c) Five spot
    d) None of the above

31. Alternating lines of producers and injectors and can be either direct or staggered is called
    a) Seven spot
    b) line drive
    c) Inverted seven spot
    d) Four spot

32. Which technique utilizes heat to make more fluid recovery from heavy oil reservoir?
    a) Thermal recovery
    b) Water flooding
    c) Chemical Recovery
    d) None of the above

33. In _____ both injection and production wells are injected with superheated steam.
    a) Steamflood
    b) Water flood
    c) Polymer flood
    d) None of the above

34. The recovery percentage for steamflood is
    a) up to 65% OOIP
    b) up to 20 % OOIP
    c) up to 15% OOIP
    d) up to 25% OOIP

35. In ________, the flame front moves from the injection well towards the producing well.
    a) Forward combustion
    b) Reverse combustion
    c) Discontinuous combustion
    d) None of the above

36. In the reverse combustion flame front moves in the ________ direction as that of injected well
    a) Same
    b) Opposite
    c) Both directions
    d) None of the above

37. In __________combustion only air is injected to maintain combustion at the front:
    a) dry forward combustion
    b) wet combustion

c) both (a) and (b)

d) None of the above

38. Which of the following parameters are necessary for in-situ combustion?

a) Oil content, oil gravity and oil viscosity

b) Pay thickness, depth and size of reservoir

c) Reservoir permeability

d) All of the above

39. The ratio of cross-sectional area contacted by displacing agent to total cross-sectional area is called

a) Vertical sweep efficiency

b) Areal sweep efficiency

c) Displacement sweep efficiency

d) None of the above

40. The area contacted by displacing agent to total area is called

a) Areal sweep efficiency

b) Vertical sweep efficiency

c) Volumetric sweep efficiency

d) None of the above

41. The ratio of volume of oil contacted by displacing agent to the volume of oil originally in place (OOIP) is called

a) Areal sweep efficiency

b) Vertical sweep efficiency

c) Volumetric sweep efficiency

d) None of the above

42. Gravity tonguing is important in

a) steam flooding operations

b) Water flooding operations

c) Both (a) and (b)

d) None of the above

43. The volumetric sweep efficiency is the product of

a) Areal efficiency and vertical sweep efficiency

b) Displacement sweep efficiency and areal sweep efficiency

c) Displacement sweep efficiency and vertical sweep efficiency

d) None of the above

44. What is the determining factor as to when a waterflood operation reaches its effectiveness?

a) Breakthrough

b) Mobility ratio

c) Sweep efficiency

d) None of the above

45. The recovery efficiency of a tertiary micellar polymer flood is the product of

a) volumetric sweep efficiency, displacement efficiency

b) areal efficiency, displacement efficiency

c) volumetric sweep efficiency, displacement efficiency, and mobility buffer efficiency

d) None of the above

46. Displacement efficiency of a micellar flood is

a) Ratio of ultimate volume of oil displaced & the volume of oil contacted

b) Product of ultimate volume of oil displaced & the volume of oil contacted

c) sum of ultimate volume of oil displaced & the volume of oil contacted

d) None of the above

47. Mobility buffer efficiency is given by:

a) $E_{MB} = (1 - E_{MBe})\left[1 - \exp\left(\frac{-0.4t_{DMB}}{E_V^{1.2}}\right)\right] + E_{MBe}$

where E_{MBe} is the mobility buffer efficiency extrapolated to $t_{DMB} = 0$ and t_{DMB} is the mobility buffer volume

b) $E_{MB} = (1 + E_{MBe})\left[1 - \exp\left(\frac{-0.4t_{DMB}}{E_V^{1.2}}\right)\right] + E_{MBe}$

where E_{MBe} is the mobility buffer efficiency extrapolated to $t_{DMB} = 0$ and t_{DMB} is the mobility buffer volume

c) $E_{MB} = (1 - E_{MBe})\left[1 + \exp\left(\frac{-0.4t_{DMB}}{E_V^{1.2}}\right)\right] + E_{MBe}$

where E_{MBe} is the mobility buffer efficiency extrapolated to $t_{DMB} = 0$ and t_{DMB} is the mobility buffer volume

d) $E_{MB} = (1 - E_{MBe})\left[1 - \exp\left(\frac{-0.4t_{DMB}}{E_V^{1.2}}\right)\right] \times E_{MBe}$

where E_{MBe} is the mobility buffer efficiency extrapolated to $t_{DMB} = 0$ and t_{DMB} is the mobility buffer volume

48. Displacement efficiency is expressed as:

a) $E_D = 1 - \frac{S'_{2r}}{S_{2r}}$; where S'_{2r} & S_{2r} are the residual oil saturation to an MP and a waterflood, respectively.

b) $E_D = 1 \times \frac{S'_{2r}}{S_{2r}}$; where S'_{2r} & S_{2r} are the residual oil saturation to an MP and a waterflood, respectively.

c) $E_D = 1 + \frac{S'_{2r}}{S_{2r}}$; where S'_{2r} & S_{2r} are the residual oil saturation to an MP and a waterflood, respectively.

d) None of the above

49. The oil recovery is given by the

a) Product of displacement efficiency, vertical efficiency and oil in place

b) Sum of displacement efficiency, vertical efficiency and oil in place

c) Ratio of displacement efficiency & vertical efficiency

d) None of the above

50. The mobility ratio is defined as

a) The ratio of mobility of displacing fluid and mobility of displaced fluid

b) The ratio of mobility of displaced fluid and mobility of displacing fluid

c) The ratio of density of displacing fluid and density of displaced fluid

d) The ratio of viscosity of displaced fluid and viscosity of displacing fluid

51. For water floods mobility ratio is given by

a) $\frac{k_{ro}\mu_o}{k_{rw}\mu_w}$; where k_{rw} and k_{ro} are the relative permeabilities to water and oil & μ_w ,and μ_o are the water and oil viscosities.

b) $\frac{k_{rw}\mu_o}{k_{r0}\mu_w}$; where k_{rw} and k_{ro} are the relative permeabilities to water and oil & μ_w ,and μ_o are the water and oil viscosities.

c) $\frac{k_{rw}\mu_w}{k_{r0}\mu_o}$; where k_{rw} and k_{ro} are the relative permeabilities to water and oil & μ_w ,and μ_o are the water and oil viscosities.

d) None of the above

52. Capillary number is the ratio of
    a) Viscous to inertial forces
    b) Viscous to gravity forces
    c) Inertial to viscous forces
    d) None of the above

53. Capillary number is given by the following expression:
    a) $N_c = \frac{v\mu_w}{\sigma_{ow}}$; where v is interstitial velocity of water, μ_w is viscosity of water, is interfacial tension between oil and water
    b) $N_c = \frac{\sigma_{ow\times}\mu_w}{v}$; where v is interstitial velocity of water, μ_w is viscosity of water, is interfacial tension between oil and water
    c) $N_c = \frac{\mu_w}{\sigma_{ow\times}v}$; where v is interstitial velocity of water, μ_w is viscosity of water, is interfacial tension between oil and water
    d) None of the above

54. Which caustic chemical is used in alkaline flooding?
    a) Sodium hydroxide
    b) Sodium silicate
    c) Sodium carbonate
    d) All of the above

55. Alkali flooding can be applied to oils whose API gravity ranges between
    a) 13° to 35°
    b) 23° to 35°
    c) 33° to 45°
    d) 3° to 35°

56. Micellar polymer is also known as
    a) Microemulsion flooding
    b) Low tension flooding
    c) Chemical flooding
    d) All of the above

57. Which of the following mechanism is involved in the oil recovery by hydrocarbon miscible flooding?
    a) Miscibility generation
    b) Increasing the oil volume
    c) Decreasing the viscosity of oil
    d) All of the above

58. CO_2 flooding method is used for
    a) heavy crude oil
    b) Waxy crude oil
    c) Medium and light crude oil
    d) None of the above

59. Cyclic steam injection uses only one well which functions as
    a) Injection well
    b) Production well
    c) Either injection and production well
    d) None of the above

60. Microbial enhance oil recovery is used for
    a) High temperature well
    b) Very low temperature well
    c) Moderate temperature well
    d) None of the above

61. Which of the following is a chemical flooding simulator?
    a) UTCHEM
    b) Kingdom
    c) Petrel
    d) None of the above

62. Which of the following statement is true?
    a) Water is relatively easy to inject into the formation than the other fluids used for secondary and tertiary recovery
    b) Water easily spreads to the oil formation than the other injected fluids to improve the oil recovery
    c) Both (a) and (b)
    d) None of the above

63. Which of the following is true about benefits of water flooding?
    a) Water is easily available
    b) Water is efficient displacing agent for light to medium gravity oil
    c) Water flooding requires low CAPEX and OPEX
    d) All of the above

64. What are the reservoir parameters which are beneficial for steam injection?
    a) Oil gravity should be above 20°API
    b) Oil viscosity should be in between 100 to 1000 cP at reservoir temperature
    c) The permeability of the formation should be above 50 mD and porosity should be above 25%
    d) All of the above

65. Which of the following are the limitations of steam flooding?
    a) Steam flooding is not normally used in carbonate reservoir
    b) A low percentage of water sensitive clay is desired for good Injectivity
    c) Both (a) and (b)
    d) None of the above

66. What are the problems of steam flooding?
    a) Adverse mobility problems
    b) Channeling of steam
    c) Gravity override
    d) All of the above

67. What are the limitations of in-situ combustion?
    a) Oil saturation and porosity must be high to minimize heat loss to the rock

b) The quantity of air required to sustain combustion should be high

c) Sufficient coke should be deposited from the oil being burned

d) All of the above

68. What is the main problem of in-situ combustion?

a) It has adverse mobility ratio

b) Produced flue gases may cause environmental problems

c) Pipe failures in the producing well as a result of very high temperature

d) All of the above

69. Which of the following is true about the selection of EOR processes?

a) Polymer flooding should be used up to about viscosities of 1000 cp

b) Conventional thermal EOR should be used for oil viscosities for more than 1000 cp

c) Both (a) and (b)

d) None of the above

70. Which of the following is true about the gas injection?

a) Nitrogen and flue gas provide gas drive

b) It enhances gravity drainage

c) Both (a) and (b)

d) None of the above

71. Chemical EOR process should be selected when

a) Crude oil gravity is more than 15°API

b) Crude oil viscosity should be less than 200 cp

c) Average permeability of formation should be more than 10mD

d) All of the above

72. Gas based EOR process may be used when

a) Crude oil gravity should be more than 20°API

b) Viscosity of oil should be $<$ 5cp

c) Average permeability should be more than 5mD

d) All of the above

73. Gas based immiscible EOR process should be used when

a) API gravity of oil should be more than 15°API

b) Viscosity of oil should be less than 25 cp

c) Average permeability should be more than 15 mD

d) All of the above

74. Gas based miscible EOR process should be used when

a) API gravity of the crude oil should be more than 30°API

b) Viscosity of the crude oil should be less than 3 cp

c) Average permeability of the formation should be more than 5mD

d) All of the above

75. Which of the following is about water and gas (WAR) injection process
    a) Combined water and gas injection improves displacement efficiency of gas flooding with improved macroscopic sweep by water injection
    b) Recovery of oil is improved by exploiting the segregation of gas to the top and accumulation of water towards the bottom
    c) Both (a) and (b)
    d) None of the above

76. Which of the following is true about tapering?
    a) It is the decrease in gas to water ratio as the flood progresses
    b) It is generally done to control gas mobility and channeling
    c) It prevents early breakthrough of gas
    d) All of the above

77. Which of the following is true about simultaneous water and gas injection (SWAR)
    a) In SWAR process, water and gas injection take place simultaneously
    b) In SWAR process, water, oil & gas injection take place simultaneously
    c) Both (a) and (b)
    d) None of the above

78. Which of the following is true about the microbial EOR process?
    a) Microbes do not consume large quantity of energy
    b) The use of microbes is not dependent on the price of crude oil
    c) Microbes can be produced rapidly from inexpensive and/or renewable resources hence microbial EOR process is potential to be more cost effective than other EOR process
    d) All of the above

Statement for linked answer questions for 79-95

The following laboratory data and field data are given for the evaluation of effectiveness of in-situ combustion process in a viscous oil reservoir

Details of combustion tube:

Combustion tube internal diameter: 0.333 ft

Length of burned pack: 6.0 ft

Porosity of pack: 35%

Volume of produced gas (dry basis): 190 SCF

Composition of injected air:

N_2: 79%

O_2 : 21%

Composition of produced gas

N_2: 84.2

O_2: 1.1

CO_2: 11.7

CO: 3.0

Oil Field data:

Pattern area, 5 acre

Formation thickness, 80 ft

Reservoir temperature, 70° F

Production well bottom hole pressure, 55 psia

Porosity, 18%

Absolute permeability, 440 md

Oil saturation, 79.9%

Production and injection well radius, 0.375 ft

Volumetric sweep of burned zone, 40%

79. Volume of nitrogen and oxygen injected in the laboratory run?
    a) 160 SCF and 42.5 SCF
    b) 260 SCF and 42.5 SCF
    c) 360 SCF and 42.5 SCF
    d) 460 SCF and 42.5 SCF

80. Volume of carbon dioxide, carbon monoxide, and unreacted oxygen produced in the laboratory run is
    a) 42.23 SCF, 5.7 SCF, 2.09 SCF
    b) 22.23 SCF, 5.7 SCF, 2.09 SCF
    c) 32.23 SCF, 5.9 SCF, 3.02 SCF
    d) 62.23 SCF, 6.7 SCF, 5.02 SCF

81. Carbon in fuel burned, hydrogen in fuel burned and weight of water formed in pounds in the laboratory run will be
    a) 0.88433 lb, 0.1618 lb, 1.46 lbs of H_2O
    b) 3.88433 lb, 0.1618 lb, 1.46 lbs of H_2O
    c) 1.88433 lb, 0.1618 lb, 1.46 lbs of H_2O
    d) 0.88433 lb, 1.1618 lb, 2.46 lbs of H_2O

82. Weight of total fuel consumed in the laboratory run.
    a) 1.04613 lbs
    b) 2.04163 lbs
    c) 3.0146 lbs
    d) None of the above

83. Fuel consumed per cubic foot of the sand pack burned and per acre-ft of the reservoir of pattern.
    a) 0.524 ft^3, 110015.14 lbs/acre-ft
    b) 5.005 lbs, 112252.0 lbs/acre-ft
    c) 7.5 lbs, 132254.0 lbs/acre-ft
    d) None of the above

84. Air injected per acre-ft of reservoir burned will be:
    a) 21.3 MMSCF
    b) 31.3 MMSCF
    c) 41.3 MMSCF
    d) 51.3 MMSCF

85. Total air required for five spot pattern will be:
    a) 5300 MMSCF
    b) 5500 MMSCF
    c) 7500 MMSCF
    d) 8500 MMSCF

86. Air flux for a minimum burning front advance of 0.125 ft/day is:
    a) 91.1 SCF/ft^2-day
    b) 61.1 SCF/ft^2-day
    c) 81.1 SCF/ft^2-day
    d) 71.1 SCF/ft^2-day

87. Maximum daily air injection rate for the field pattern and the time required to reach this maximum rate will be:
    a) 7.68 MMSCF/D, 125 days
    b) 9.68 MMSCF/D, 125 days
    c) 17.68 MMSCF/D, 125 days
    d) 7.68 MMSCF/D, 225 days

88. Volume of air injected while reaching maximum air injection rate, and air injected during constant rate period are:

a) 580 MMSCF, 4340 MMSCF

b) 480 MMSCF, 4340 MMSCF

c) 480 MMSCF, 4640 MMSCF

d) 680 MMSCF, 4340 MMSCF

89. Duration of constant rate period and the time to complete the entire burning operation are:

a) 765 days, 815 days

b) 565 days, 915 days

c) 565 days, 815 days

d) 865 days, 815 days

90. Maximum air injection pressure required for an assumed air viscosity of 0.0186 cp and a permeability to injection air of 25 md is:

a) 766 psia

b) 566 psia

c) 666 psia

d) 866 psia

91. Oil displaced from the reservoir burned, oil displaced from the unburned reservoir, total oil recovery, and overall oil recovery efficiency is:

a) 803 bbl/acre-ft, 546.0 bbl/acre-ft, 554 bbl/acre-ft, 49.5%

b) 803 bbl/acre-ft, 446.0 bbl/acre-ft, 554 bbl/acre-ft, 49.5%

c) 803 bbl/acre-ft, 446.0 bbl/acre-ft, 554 bbl/acre-ft, 69.5%

d) 103 bbl/acre-ft, 446.0 bbl/acre-ft, 554 bbl/acre-ft, 49.5%

92. Barrels of water produced per acre-ft is:

a) 316 bbl/acre-ft

b) 516 bbl/acre-ft

c) 216 bbl/acre-ft

d) 816 bbl/acre-ft

93. Oil recovery per MMSCF of air injected is:

a) 81.9 bbl/MMSCF

b) 51.9 bbl/MMSCF

c) 61.9 bbl/MMSCF

d) 41.9 bbl/MMSCF

94. Maximum oil producing rate from the pattern is:

a) 422 BOPD

b) 522 BOPD

c) 322 BOPD

d) 622 BOPD

95. Barrels of water produced per MMSCF of air injected and the maximum water-producing rate is:

a) 16.13 barrels/MMSCF, 324 Barrels of water/day

b) 26.13 barrels/MMSCF, 124 Barrels of water/day

c) 16.13 barrels/MMSCF, 124 Barrels of water/day

d) 16.13 barrels/MMSCF, 524 Barrels of water/day

96. What is primary producing mechanism on reservoir pressure and recovery efficiency of an oil well?

a) Liquid and rock expansion

b) Gas cap expansion

c) Water influx

d) All of the above

97. Which of the following is true about liquid and rock expansion mechanism?

a) It declines reservoir pressure rapidly and continuously

b) GOR remains low and constant

c) The recovery efficiency of this production mechanism varies from 1 to 10%

d) All of the above

98. Which of the following is true about solution gas drive mechanism?

a) It declines reservoir pressure rapidly and continuously

b) Initially low GOR then GOR rises to maximum and then drops

c) Recovery efficiency of this mechanism varies from 5 to 35%

d) All of the above

99. Which of the following is best short term method for improved oil recovery?

a) Infill Drilling

b) WAG

c) FAWAG

d) CO_2 Injection

100. Which of the following statement is true about gravity drainage mechanism:

a) It declines reservoir pressure rapidly and continuously

b) Water production is absent or negligible

c) Recovery efficiency of this producing mechanism varies from 40-80%

d) All of the above

Answers

Q1.	a	Q21.	a	Q41.	c	Q61.	a	Q81.	a
Q2.	b	Q22.	b	Q42.	a	Q62.	c	Q82.	a
Q3.	a	Q23.	a	Q43.	a	Q63.	d	Q83.	a
Q4.	a	Q24.	a	Q44.	a	Q64.	d	Q84.	a
Q5.	a	Q25.	a	Q45.	c	Q65.	c	Q85.	a
Q6.	a	Q26.	b	Q46.	a	Q66.	d	Q86.	b
Q7.	a	Q27.	a	Q47.	a	Q67.	d	Q87.	a
Q8.	b	Q28.	c	Q48.	a	Q68.	d	Q88.	b
Q9.	b	Q29.	c	Q49.	a	Q69.	c	Q89.	c
Q10.	a	Q30.	a	Q50.	a	Q70.	c	Q90.	c
Q11.	d	Q31.	b	Q51.	b	Q71.	d	Q91.	b
Q12.	a	Q32.	a	Q52.	a	Q72.	d	Q92.	c
Q13.	a	Q33.	a	Q53.	a	Q73.	c	Q93.	d
Q14.	b	Q34.	a	Q54.	d	Q74.	d	Q94.	c
Q15.	d	Q35.	a	Q55.	a	Q75.	d	Q95.	c
Q16.	a	Q36.	b	Q56.	d	Q76.	d	Q96.	d
Q17.	b	Q37.	a	Q57.	d	Q77.	a	Q97.	d
Q18.	a	Q38.	d	Q58.	c	Q78.	d	Q98.	d
Q19.	d	Q39.	a	Q59.	c	Q79.	a	Q99.	a
Q20.	a	Q40.	a	Q60.	c	Q80.	b	Q100.	d

CHAPTER **11**

Unconventional Hydrocarbon Resources

1. Which of the following is true about the mechanism for gas flow in the coal?
   a) Desorption of gas from the coal surface inside the micropores
   b) Diffusion of gas through the micropores
   c) Darcy flow through the fractures
   d) All of these

2. Coal has a
   a) single porosity system
   b) dual porosity system
   c) triple porosity system
   d) None of the above

3. Diffusion through coals micropores may be due to
   a) Bulk diffusion
   b) Kmudsen type diffusion flow and Surface diffusion
   c) Combination of all the above diffusion process
   d) All of these

4. Sorption time is defined as the time required to recover _______ of methane on a sample of coal characterized the diffusion of methane through the coal matrix and into the cleat system.
   a) 25%
   b) 63%
   c) 50%
   d) 40%

5. Coal bed methane is
   a) Conventional source of energy
   b) Unconventional source of energy
   c) Renewable source of energy
   d) None of the above

6. The gas production in coal occurs
   a) By flow of free gas from natural fractures
   b) Desorption of gas from cleat surfaces
   c) Diffusion of gas through coal matix
   d) All of these

7. Coal desorption characteristics and isotherm are function of
   a) Coal properties of rank
   b) Cleat concentration
   c) Ash and moisture content
   d) All of these

8. Coal reservoir properties that control fluid flow are
   a) Fracture permeability and orientation
   b) Coal matrix diffusion characteristics
   c) Desorption isotherm behavior
   d) All of the above

9. Which of the following log is suited in coal bed methane wells?
   a) Resistivity log
   b) Gamma ray log
   c) Neutron log, and caliper log
   d) All of these

10. The density log provides a means of determining__
    a) Ash content of coal
    b) Estimate of gas content
    c) Both (a) and (b)
    d) None of thes

11. Permeability of coal can be estimated by
    a) Microresistivity log
    b) Density log
    c) Gamma ray log
    d) None of the above

12. Coal rank is influenced by:
    a) Depth
    b) Length of burial
    c) Temperature
    d) All of these

13. In coal seam, how much percentage of gas is stored by physical adsorption on molecular structure of the coal surface?
    a) more than 95%
    b) More than 75%
    c) More than 50%
    d) None of the above

14. Gas content of coal and associate desorption isotherm can be determined by
    a) Core analysis
    b) Mud analysis
    c) SP log analysis
    d) None of the above

15. Isotherm is used to determine
    a) Suction capacity of coal
    b) At reservoir condition
    c) Moisture content of coal
    d) Rank of the coal

16. The coal desorption characteristics and isotherm are the function of
    a) Coal rank
    b) Cleat concentration
    c) Ash and moisture content
    d) All of these

17. ______ is the process by which methane molecules detach from the microscopic surface of the coal matrix and enter the cleat system where they exist as a free gas.
    a) Diffusion
    b) Desorption
    c) Adsorption
    d) None of the above

18. Important characteristics of the production history of many coal-bed methane wells include:
    a) Initial large and declining water cuts and low and improving gas production rates
    b) Initial large but declining gas/liquid ratios and low but increasing water production rates
    c) Stabilized gas/liquid ratios and decreasing gas rates
    d) Stabilized gas/liquid ratios and increasing gas and water production rates

19. ________ is the maximum capacity of ash in moisture free coal as determined in laboratory by measuring adsorbed gas in crushed coal sample at increasing pressure.
    a) Langmuir volume constant
    b) Langmuir pressure constant
    c) Both of these
    d) None of the above

20. Langmuir pressure is the pressure at which
    a) gas storage capacity of coal is equal to ½ of the Langmuir volume
    b) gas storage capacity of coal is equal to of the Langmuir volume
    c) gas storage capacity of coal is double of the Langmuir volume
    d) None of the above

21. Langmuir volume is the volume of gas at
    a) Zero pressure
    b) Atmospheric pressure
    c) Infinite pressure
    d) None of the above

22. Which of the following statement is false?
    a) Langmuir pressure is defined as a pressure at which 75% of gas is desorbed
    b) In shale formations, gas velocity in fractures is very high and the flow pattern deviates from Darcy flow
    c) Klinkenberg effect is very significant and might not be ignored while modeling shale gas flow
    d) None of the above

23. Which of the following statement is one of the assumptions of Langmuir equation?
    a) One gas molecule is adsorbed at a single adsorption site
    b) An adsorbed molecule don not affect the molecules on the neighboring site
    c) Adsorption is on open surface and there is no resistance to gas access to adsorption sites
    d) All of these

24. Diffusion of gas in coal can be described using Fick's law and it is the function of
    a) Matrix gas concentration
    b) Cleat spacing
    c) Diffusion constant
    d) All of these

25. Which of the following method is one of the proposed methods for gas recovery from hydrates?
    a) Depressurization
    b) Thermal methods
    c) Inhibitor injection
    d) All of these

26. In the thermal methods of gas recovery the gas hydrate may be dissociated by
    a) Heating the reservoir below the hydrate formation temperature
    b) Heating the hydrate beyond hydrate formation temperature
    c) Both (a) and (b)
    d) None of the above

27. The depressurization method of gas recovery from hydrates usually deal with dissociating in situ gas hydrates by
    a) Decreasing reservoir pressure below hydrate equilibrium
    b) Decreasing reservoir pressure above hydrate equilibrium
    c) Both (a) and (b)
    d) None of the above

28. The inhibitor injection method of gas recovery from hydrates usually deal with dissociating gas hydrates by injecting_____ in to the reservoir to decrease hydrate stability conditions.
    a) Methanol
    b) Glycol
    c) Both of these
    d) None of the above

29. Where does gas hydrate occur?
    a) in deep ocean sediments
    b) in permafrost
    c) Both (a) and b)
    d) None of the above

30. Which gas hydrates have the most economic potential?
    a) permafrost hydrates
    b) deep ocean sediment gas hydrates
    c) Both (a) and (b)
    d) None of the above

31. Which of the following is a type of cleat?
    a) Face
    b) But
    c) Both (a) and (b)
    d) None of the above

32. ______ shift the hydrate equilibrium curve toward higher pressures and lower temperatures by changing the activity of water molecules.
    a) thermodynamic hydrate inhibitors
    b) thermodynamic hydrate promoters
    c) Both (a) and (b)
    d) None of the above

33. Which of the following is thermodynamic hydrate inhibitor?
    a) Methanol
    b) Ethylene Glycol
    c) Both (a) and (b)
    d) None of the above

34. The conditions for the gas hydrate formations are
    a) High pressure and low temperature
    b) Low Pressure and High temperature
    c) High pressure and High Temperature
    d) Low pressure and low temperature

35. The advantage of using antiagglomerate type hydrate inhibitor is
    a) It can handle a higher degree of sub-cooling
    b) It can sustain a longer shut-in period.
    c) Both (a) and (b)
    d) None of the above

36. The common problems associated with the formation of hydrates in oil and gas facilities are:
    a) Reduction in the gas throughput
    b) plugging of flowlines,
    c) damage to downstream equipment and increased safety concerns
    d) All of the above

37. The presence of these acid gases shifts the
    a) hydrate equilibrium curve to the right
    b) hydrate equilibrium curve to the left
    c) Both (a) and (b)
    d) None of the above

38. Which of the following is true about the oil shale?
    a) Organic rich shale that is old enough, but have never been buried deep (i.e.hot) enough to enter the oil window.
    b) Organic matter is still preserved in the form of kerogen
    c) Crude oil can be produced from oil shale by heating it up, but the production is costly
    d) All of these

39. Shales are fine grained clastic sedimentary rock made up of
    a) Clay
    b) Silt size particles
    c) Both (a) and (b)
    d) None of the above

40. India has huge shale potential in________
    a) Cambay Basin
    b) Cauvery Basin
    c) Damodar Valley Sub Basin
    d) All of these

41. Hydrocarbon generative potential of shale depends on
    a) Amount of organic matter present in it
    b) Type of organic matter present in it
    c) Maturity of organic matter present in it
    d) All of these

42. Major technologies required for shale gas exploitation are
    a) Horizontal drilling and Multilateral drilling
    b) Pad drilling and Stacked well
    c) Hydraulic fracture stimulation
    d) All of these

43. Which of the following is true about the porosity of shale gas reservoirs?
    a) Porosity of shale gas reservoir ranges from 1% to 5% before fracturing
    b) Porosity of shale gas reservoir is about 11% after fracturing
    c) Both (a) and (b)
    d) None of the above

44. Which of the following is true about the permeability of shale gas reservoir?
    a) The permeability of the shale gas reservoir is less than 10^{-6} darcy before fracturing
    b) The permeability of the shale gas reservoir is around 10^{-3} darcy after fracturing
    c) Both (a) and (b)
    d) None of the above

45. Shale gas formation exhibits
    a) Single porosity matrix
    b) Dual porosity matrix
    c) Triple porosity matrix
    d) None of the above

46. How many flow regimes are defined for the flow of natural gas from shale gas reservoirs?
    a) One
    b) Two
    c) Three
    d) Four

47. In the very early time flow period , the flow of shale gas occurs from
    a) Hydraulic fractures
    b) Matrix
    c) Matrix and fractures
    d) None of the above

48. In the intermediate time flow period , the flow of shale gas occurs from
    a) Hydraulic fractures
    b) Matrix
    c) Matrix and fractures
    d) None of the above

49. In the late time flow period , the flow of shale gas occurs from
    a) Hydraulic fractures
    b) Matrix
    c) Matrix and fractures
    d) None of the above

50. The proposed methods for enhancing shale gas recovery are
    a) CO_2 injection and enhanced CH_4 recovery
    b) High temperature supercritical carbon dioxide flooding
    c) Both (a) and (b)
    d) None of the above

51. Multistage hydraulic fracturing can be accomplished by
    a) Plug and Perf method
    b) Sliding sleeve method
    c) Both (a) and (b)
    d) None of the above

52. At what depth, Nitrogen foam will be suitable for fracturing of shale gas formation
    a) Depth should be less than 6000 ft
    b) Depth should be more than 6000 ft
    c) Depth should be more than 10000 ft
    d) None of the above

53. If reservoir pressure gradient is less than 0.433 psi/ft the most suitable fracturing fluid will be
    a) Nitrogen foam
    b) Slick water
    c) CO_2 foam
    d) None of the above

54. At what depth, slickwater will be suitable for fracturing of shale gas formation
    a) The depth should be less than 6000 feet
    b) The depth should be more than 17000 ft
    c) Depth should be in between 6000 ft and 17000 ft
    d) None of the above

55. If reservoir pressure gradient of the shale gas reservoir is more than 0.433 psi/ft and depth is more than 6000 ft the most suitable fracturing fluid will be
    a) Nitrogen foam
    b) Slick water
    c) CO_2
    d) None of the above

56. The following properties of shale gas formations are given:
    Depth = 9000 ft
    Thickness = 200 ft
    TOC = 9.0%
    Gas content = 100 scf/ton
    Quartz content = 63%
    Clay content = 30%
    Reservoir pressure gradient = 0.52 psi/ft
    Adsorbed gas = 23%
    Which fracturing fluid should be selected for the fracturing of above shale gas formation?
    a) Slickwater
    b) Nitrogen foam
    c) Carbon dioxide
    d) None of the above

57. Shales having high clay content cause_______ during drilling.
    a) Hole enlargement problem
    b) Sloughing shale problem
    c) Pipe sticking
    d) All of these

58. The preferred drilling fluid for the drilling of shale formations are
    a) Amine based drilling fluids
    b) KCL/PHPA water based drilling fluid system
    c) Silicate based drilling fluid system
    d) All of these

59. Factor which decides well completion for shale gas formations are
    a) Total Organic Content and Gas Content
    b) Clay Content, and Quartz content
    c) thickness and reservoir pressure
    d) All of these

60. Gas shale is ________
    a) source rock
    b) reservoir rock
    c) seal
    d) Combination of source rock, reservoir rock and seal

61. Areas near a natural fault are not good for shale gas production because_____
    a) some of the gas has escaped along the fault
    b) during hydraulic frac the fault takes most of the frac fluid
    c) Both (a) and (b)
    d) None of the above

62. What is the major concern for shale gas production?
    a) volume of water used during hydraulic frac
    b) the flow back, and hydraulic frac penetrating and polluting freshwater reservoirs
    c) Both (a) and (b)
    d) None of the above

63. Which are considered to be best gas shales?
    a) Relatively shallow depth gas shales
    b) Shales having high organic carbon content and high concentration of natural fractures
    c) Both (a) and (b)
    d) None of the above

64. Which of the following is true about oil shales?
    a) Oil shales are sedimentary rocks in a type of organic matter called kerogen
    b) Oil shale is heated to about 660° and then kerogen is transformed into shale oil
    c) Both (a) and (b)
    d) None of the above

65. The preferred drilling fluid for the drilling of shale formations are
    a) Amine based drilling fluids
    b) KCL/PHPA water based drilling fluids
    c) Silicate based drilling fluids
    d) All of these

66. Which of the following is true about tar sands?
    a) It is composed of bitumen, water and sediments
    b) surface deposit of tar sands are typically 130 to 200 ft
    c) Both (a) and (b)
    d) None of the above

67. Which of the following is true about oil sands?
    a) Oil sand is unconventional hydrocarbon resource

b) Oil sands are thick slurry composed of water, sand and and bitumen

c) Both (a) and (b)

d) None of the above

68. Which of the following statement is false?

a) 80% of the oil sands are buried too deep for surface mining

b) oil sands are similar to heavy crude oil

c) What are flowing, viscous deposits of bitumen mixed with sand, water and clay -oil sands

d) None of the above

69. Oil sands within how many meters of the surface can be surface mined?

a) 75m

b) 100 m

c) 300m

d) 500 m

70. Which of the following method is used to stimulate unconventional reservoir for more recovery of unconventional natural gas?

a) Hydraulic Fracturing

b) Matrix Acidization

c) Acid Washing

d) None of the above

71. What is the main advantage of hydraulic fracturing?

a) It increases flow rate by greatly increasing the surface are of shale exposed to the wellbore

b) Conducted at depth of 5,000-10,000 feet

c) Both (a) and (b)

d) None of the above

72. Which of the following fracturing fluid is used for unconventional reservoirs?

a) Foam based fracturing fluid

b) Oil based fracturing fluid

c) Both (a) and (b)

d) None of the above

73. Oil sands below the 75m threshold have to be extracted using which method?

a) In-situ

b) Surface mining

c) Both (a) and (b)

d) None of the above

74. Which of the following method is the in-situ extraction method of oil sand?

a) Steam Assisted Gravity Drainage

b) Cyclic Steam Stimulation

c) Solvent

d) All of these

75. Which of the following is true about Steam Assisted Gravity Drainage?

a) It is most popular In Situ method for the extraction of oil sand

b) In this method, steam is pumped through horizontal well to liquefy the bitumen and liquefied bitumen is pumped to the surface through a second well

c) It is very energy expansive process

d) All of these

76. Which of the following is true about cyclic steam stimulation process?

a) It is an important in-situ extraction method of oil sand

b) In this method, steam is pumped down a vertical well to soak or liquefy bitumen and liquefied bitumen is pumped to the surface through the same well.

c) Both (a) and (b)

d) None of the above

77. Which of the following is true about the solvent extraction?

a) In this method, solvent is pumped underground through horizontal well to liquefy the bitumen

b) Solvent extraction is 90% more energy efficient than other in situ methods

c) Both (a) and (b)

d) None of the above

78. Methane hydrates form under conditions of:

a) -15 to 15 degree celsius and High Pressure

b) 15 to 25 degree and high pressure

c) -15 to 15 degree Celsius and low Pressure

d) 15 to 25 degree Celsius and high pressure

79. Heavy bitumen can be upgraded to a light crude oil and this upgraded oil is called

a) Synthetic crude oil

b) Synthetic petrol

c) Synthetic diesel

d) None of the above

80. Oil sand resources are found as slight less viscous oil in some areas and non steam production methods may be used for those resources. Which of the following is suitable non steam production method?

a) CHOPS

b) Polymer Flooding

c) Both (a) and (b)

d) None of the above

81. Which of the following artificial lift technique is used in SAGD operations?

a) Gas lift technique

b) Electrical submersible pumping system

c) Both (a) and (b)

d) None of the above

82. Which of the following is true about cyclic steam stimulation?

a) It was the first process used to commercially recover oil sand in situ

b) It is also called huff and puff method

c) It is a three stage steam injection process that uses vertical, deviated and horizontal wells

d) All of these

83. The oil having API gravity 14-21 are classified as
    a) Heavy crude oil
    b) Extra heavy crude oil
    c) Medium heavy crude oil
    d) Bitumen

84. The oil having API gravity <10 are classified as
    a) Heavy crude oil
    b) Extra heavy crude oil
    c) Medium heavy crude oil
    d) Bitumen

85. The oil having API gravity 21-31 are classified as
    a) Heavy crude oil
    b) Extra heavy crude oil
    c) Medium heavy crude oil
    d) Bitumen

86. The oil having API gravity 10-14 are classified as
    a) Light crude oil
    b) Heavy crude oil
    c) Extra heavy crude oil
    d) Medium heavy crude oil

87. Crude oils having API gravity lower than 22 degrees are termed as _____
    a) Unconventional resources
    b) Conventional resources
    c) Both (a) and (b)
    d) None of the above

88. As per world petroleum conference, heavy oil are those having API gravity
    a) Below 22.3 degrees
    b) Below 24.3 degrees
    c) Below 25.3 degrees
    d) None of the above

89. As per American Petroleum Institute, heavy oil is defined as those having API gravity equal to or smaller than
    a) 20 degrees
    b) 22 degrees
    c) 24 degrees
    d) None of the above

90. As per Petrobras, heavy oils are those having API gravity in the range
    a) Between 10-19 °API
    b) Between 5-15 °API
    c) Between 10-20 °API
    d) None of the above

91. As per National Petroleum Institute of Brazil, heavy oils are those having API gravity in the range
    a) Between 10-20 °API
    b) Between 10-19 °API
    c) Between 5-15 °API
    d) None of the above

92. As per National Petroleum Institute of Brazil, extra heavy oils are those having API gravity
    a) Less than or equal to 12oAPI
    b) Less than or equal to 10°API
    c) Less than or equal to 15°API
    d) None of the above

93. Extra heavy crude oil exhibits viscosities of
    a) Between 10-100 centipoise
    b) Between 100-10000 centipoise
    c) Both (a) and (b)
    d) None of the above

94. Which of the following is the challenge associated with production of heavy oil?
    a) The need for artificial lift to produce viscous oils from wells onshore and offshore
    b) Solid production associated with processing of heavy oil
    c) Flow assurance issues
    d) All of these

95. Which of the following EOR method is used for heavy oil?
    a) Cyclic steam stimulation method
    b) In-situ combustion method
    c) Both (a) and (b)
    d) None of the above

96. Which of the following are effective EOR methods for cold heavy oil production?
    a) Immiscible CO_2 Flooding
    b) CO_2 water alternating gas
    c) Both (a) and (b)
    d) None of the above

97. Major limitations for using polymer flooding are:
    a) Adsorption of polymer onto reservoir rock
    b) Polymer degradation
    c) Injectivity loss
    d) All of these

98. Which of the following may be used to improve the sweep efficiency of CO_2 injection
    a) Foam
    b) Polymer Gel
    c) Both (a) and (b)
    d) None of the above

99. Steam injection method may be used for the recovery of heavy oil when
    a) Oil gravity is 10-25 °API
    b) Depth is 200-5000 ft
    c) Target oil is greater than 500 bbl/acre ft
    d) All of the above

100. The partial upgrading of heavy oil or bitumen includes
    a) Hydrocracking
    b) Hydrotreating
    c) Both (a) and (b)
    d) None of the above

Answers

Q1.	d	Q21.	c	Q41.	d	Q61.	c	Q81.	c
Q2.	b	Q22.	a	Q42.	d	Q62.	c	Q82.	d
Q3.	d	Q23.	d	Q43.	c	Q63.	c	Q83.	a
Q4.	b	Q24.	d	Q44.	c	Q64.	c	Q84.	d
Q5.	b	Q25.	d	Q45.	b	Q65.	d	Q85.	c
Q6.	d	Q26.	b	Q46.	c	Q66.	c	Q86.	c
Q7.	d	Q27.	a	Q47.	a	Q67.	c	Q87.	a
Q8.	d	Q28.	c	Q48.	c	Q68.	d	Q88.	a
Q9.	d	Q29.	c	Q49.	b	Q69.	a	Q89.	a
Q10.	c	Q30.	a	Q50.	c	Q70.	a	Q90.	a
Q11.	a	Q31.	c	Q51.	b	Q71.	c	Q91.	a
Q12.	d	Q32.	a	Q52.	a	Q72.	c	Q92.	b
Q13.	a	Q33.	c	Q53.	a	Q73.	a	Q93.	b
Q14.	a	Q34.	a	Q54.	c	Q74.	d	Q94.	d
Q15.	a	Q35.	c	Q55.	b	Q75.	d	Q95.	d
Q16.	d	Q36.	d	Q56.	a	Q76.	c	Q96.	c
Q17.	b	Q37.	a	Q57.	d	Q77.	c	Q97.	d
Q18.	a	Q38.	d	Q58.	d	Q78.	a	Q98.	a
Q19.	a	Q39.	c	Q59.	d	Q79.	a	Q99.	d
Q20.	a	Q40.	d	Q60.	d	Q80.	c	Q100.	c

Hints and Solutions of Important Questions

Chapter 2: Oil and Gas Well Technology

Question. 108

$ø_{600} = 60$

Apparent Viscosity = $ø_{600}/2$ = 30 cp

Question. 109

$ø_{600}$ = 46, $ø_{300}$ = 28

Plastic viscosity (PV) = $ø_{600} - ø_{300}$ = 18 cp

Yield point = $ø_{300}$ – PV = 18 lb/ft²

Question. 110

Oil density ($\bullet_m$) = 42 °API

Oil specific gravity = 141/(°API+131.5) = 0.8156

Oil density ($\bullet_m$), ppg = 0.8156 x density of water = 0.8156 x 8.33ppg = 6.794 ppg

$(\bullet_m)_{avo}$ = $(\bullet_1 V_1 + \bullet_2 V_2)/(V_1+V_2)$

= (679.4 + 9040)/900

= 10.8 ppg

Question. 111

D = 700 ft, Mud density W_f = 7.5 lb/gal, Water density

W_i = 8.33 lb/gal

Let x be the volume of air requirement

$$\frac{x}{100-x} = \frac{2.3\log(3.53\times10^{-3}\,DW_f+1)}{3.53\times10^{-3}\,D(W_i-W_f)}$$

x = 0.2545 cuft(air)/cuft(mud)

Question. 112

Hydrostatic pressure, P(psi) = 0.052 × ρ (ppg) × D (ft)

1 gm/cc = 8.33ppg

∴ P(psi) = 1430

Question. 113

Hydrostatic pressure, P (psi) = 0.052 × ρ (ppg) × D(ft)

Hydrostatic pressure, P(psi) = 6606.6

Question. 162

W = 40 tons, n = 10

Efficiency (E) = $[k(1-k^n)]/n(1-k)]$ = 0.8109, k = 0.9615

Tension on fast line (F_f)

$= W/En = 4.93$

Tension on dead line (F_d)

$= W/n = 4$

Actual load on derrick, F

$= W + F_f + F_d$

$= 48.93$

Question. 164

$$F_f = \frac{w(1-K)}{K(1-K^n)} = 29710.6 \text{ lbs}$$

Question. 165

$F_d = W/n = 25{,}000$ lbs

Question. 166

$F = W + F_f + F_d = 254{,}710$ lbs

Question. 172

Rod size d = 2 in.

Pump inner size D = 6 in., Stroke length S = 1 8 ft

Volumetric efficiency n = 87%, Pump rate SPM = 4 8 spm

Volumetric output per minute = $9.09 \times 10^{-4} \times S(2D^2\text{-}d^2)n \times SPM$ = 8.27 bbl/min

Question. 173

Radius of curvature, R = 360/2п×1/BUR = 1910.8 ft, Buildup rate (BUR) = 3°/100ft

$\text{Tan } x = \frac{OM}{ON} = \frac{d-R}{D}$; x = -2.0467°

•ONM, Cos x = D/ON

•EOM, $\text{Sin } y = \frac{R}{ON} = \frac{RD}{\text{Cos } x}$, y

= 24.96°

• = x+y = 22.9°

At the end of buildup section, measured depth, AE

AE = BE(Curved length) +AB

AE = (•/BUR) + AB = 763.43 + 3000

= 3763.43 ft

Question. 177

• = 90° was achieved at 12,157 MD

BE (curved) = AE-AB = 12,000 ft,

R = 100 ft

BUR = •/ BE = 0.57°/100ft = 57°/100ft

57°/100ft build up rate suggests the short radius well

Question. 182

V_s – Volume of Solid

V_{m2} – volume of mud

$•_s$ – Density of solid

$•_m$ – Density of mud

Volume % solids =

$$\frac{v_s}{v_{m2}} \times 100 = \frac{r_{m2}}{r_s - r_m 1} \times 100 =$$

$$\frac{9.5 - 8.33}{(2.5 * 8.33)\, 8.33} \times 100 = 9.4\%$$

Question. 183

Weight % solids =

$$\frac{PsVs}{P_{m2}\, V_{m2}} \times 100 = \frac{Ps\,(P_{m2} - P_{m2}}{P_{m2}\,(Ps - P_{M2}} \times 100$$

$$= \frac{20.8(9.5 - 8.33}{9.5\,(20.8 - 8.33} \times 100$$

= 20.6%

Question. 184

ρs – Density of solid

ρm – Density of mud

0.03 =

$$\frac{pm2 - pm1}{ps - pm1} = \frac{pm2 - 1.0}{2.5 - 1.0}$$

•m2 = 1.045 gm/cc = 8.7 lb/gal

Question. 185

V_s – Volume of solid

V_m – Volume of mud

V_s = 1000 (1.045 -1.0) / 2.5-1.0

= 30 cc

= 2.5 X 30 = 75 gm

Also, $V_{m1} = V_{m2} - V_s$

= 1000-30

= 970 cc water

Question. 186

V_s – Volume of solid

V_m – Volume of mud

$$V_s = \frac{1000\,(10-8.7)}{35.5-10}$$

= 50.4 cc or 4.3 × 50.4 = 217 gm

V_{m2} = 1000 + 50.4

= 1050 cc

Question. 187

S_B – Sacks of barite

V_m – Volume of mud

$$SB = \frac{15 \times 100\,(14-10)}{35.5-14}$$

= 275 sacks/

$$100 \text{ bbl} = 2750 \frac{\text{Sacks}}{1000 \text{ bbl}}$$

$$Vm2 = 1000 + \frac{\text{Sacks}}{1000 \text{ bbl}}$$

= 1180 bbl

Question. 188

V_w – Volume of water

V_m – Volume of mud

(1000 bbl) (42 gal/bbl) (12 lb/gal) + (V_w) (42) (8.33) = (1000 + V_w) (42) (10)

From this V_w = 1200 bbl

V_{m2} = 1000+ 1200

= 2200 bbl

Question. 189

$$\text{CMC needed} = \frac{350}{100} \times 2.0 \times 1000$$

= 700 lb

Question. 190

V_m – Volume of mud

S – Stroke length

N – Number of strokes

D – Pump liner diameter

d – Piston rod diameter

6190 × V_m / (S ×N ($2D^2 - d^2$) × e) = 6190 × 1000 / 16 × 40 × ($2 \times 7.5^2 - 2.25^2$) ×0.90 = 100 min

Question. 191

(Note- In question 3.5 inch is missing)

$$L = 735 \times 10^3 \times \frac{13.3 \times 4 \times 12}{60{,}000}$$

= 7,800 ft

Question. 192

p_s + 10,000 (0.85) (0.433) = 1000 (14/8.33) (0.433) + 9000 (10/8.33) (0.433)

p_s = 1700 psi (surface pressure)

Question. 193

•p – Pressure Drop

$$d = \sqrt{3 \times \frac{3}{8} \times \frac{3}{8}} = 0.65 \text{ inch}$$

(equivalent single nozzle diameter)

Hence,

$$\ddot{A}p\, \frac{500 \times 500 \times 10}{7430 \times 0.95 \times 0.95 \times 0.65^4}$$

= 2100 psi

Question. 194

ECD, ppg = Annular pressure loss/0.052×D + Mud weight in use, ppg

ECD, ppg = 200 psi / 0.052 × 10,000ft + 9.6ppg = 10.0 ppg

Question. 195

Maximum allowable mud weight (ppg from leak off test data = (Leak of pressure, psi) / 0.052 × (Casing shoe TVD, ft) + mud weight

ppg = 1140psi/ 0.052 × 4000ft + 10.0 = 15.48 ppg

Question. 196

PO, gpm = [3(D^2 x 0.7854) S]0.00411 x SPM

Where, D = liner diameter, in.

S = stroke length, in.

SPM = strokes per minute

PO, gpm = [3 (7^2 x 0.7854) 12] 0.00411 x 80 = 455.5gpm

Question. 197

PO @ 100% = 0.000162 x 14 x [2 $(5.5)^2 - 2^2$] = 0.128142bbl/stk

Adjust pump output for 85% efficiency:

PO @ 85%, = 0.128142 bbl/stk x 0.85 = 0.10892 bbl/stk

Question. 198

Barrels displaced = 5 stands x 92ft/std x 0.0075bbl/ft

= 3.45

Hydrostatic pressure decrease, psi = [3.45/ (0.0773 – 0.0075)] × 0.052 × 11.5 = 29.56 psi

Question. 199

Barrels displaced = 5 stands × 92 ft × (0.01776 bbl/ft + 0.0075 bbl/ft)

= 11.6196

Hydrostatic pressure decrease, psi = [11.6196/ {(0.0773 – (0.01776+0.0075)}] × 0.052 × 11.5

= 133.52 psi

Question. 200

Feet = overbalance, psi (casing cap. – pipe disp., bbl/ft)/ (mud wt., ppg x 0.052 x pipe disp., bbl/ft)

= 150 (0.0773 – 0.0075)/ (11.5 × 0.052 × 0.0075)

= 2334

Question. 201

Feet = overbalance, psi x (casing cap. – pipe cap. – pipe disp.)/ mud wt., ppg x 0.052 x (pipe cap. + pipe disp., bbl/ft)

Feet = 150 x (0.0773 – (0.01776+0.0075))/ 11.5 x 0.052 x (0.01776 + 0.0075) = 516.8

Question. 202

New circulating pressure, psi = 1800 × (30 spm/ 60 spm)2 = 450 psi

Question. 203

Hydrostatic pressure required:
HP, psi = 12.2ppg x 0.052 x 184ft = 117 psi
Difference in pressure gradient, psi/ft:
psi/ft = (13.2 ppg – 12.2 ppg) x 0.052 = 0.052

Length of slug in drill pipe, ft:
Slug length, ft = 117psi/0.052 = 2250ft
Volume of slug, bbl:

Slug vol, bbl = 2250ft x 0.01422bbl/ft = 32.0 bbl

Question. 204

Length, ft = (WOB ×f) / (Wdc × BF)

Where WOB = desired weight to be used while drilling

f = safety factor to place neutral point in drill collars

Wdc = drill collar weight, lb/ft

BF = buoyancy factor a) Buoyancy factor (BF): 65.5 - 12.0 ppg / 65.5

BF = 0.8168 b) Length of bottomhole assembly necessary:

Length, ft = (50,000) lb × (1.15) / (147) (0.8168) = 479 ft

Question. 205

Weight of materials, lb/sk:
Weight, lb/sk = 94 + (8.33 × vol of water, gal) + (% of additive × 94)

Question. 206

Volume of slurry, gal/sk: Volume =

$$\frac{94\ \text{Ib / sk}}{SG \times 8.33} + \frac{\text{wt of additive, Ib / sk}}{2}$$

$$+\ \text{water volume, gal}$$

Question. 207

Water requirement using material balance equation:
$D_1V_1 = D_2V_2$

Weight of material, bbl/sk = 94 + (8.33 × Y) + (0.06 × 94) = 99.64 + 8.33Y

Weight of additive, lb/sk = 0.06×94.0= 5.64 lb/sk

Volume of slurry, gal/sk = 94/ (3.15×8.33) + 5.64/ (2.65×8.33) + Y= 3.86 + Y

Water requirement using material balance equation:
99.64 + 8.33Y = (3.86 + Y) ×14
Y = 8.0 gal/sk of cement

Question. 208

Amount of high density additive required per sack of cement to achieve a required cement slurry density is given by

$$X = \frac{\left\{\dfrac{\text{wt} \times 11.207}{SG_C}\right\} + (\text{wt} \times \text{cw}) - 94 - (8.33 \times \text{cw})}{\left(1 + \dfrac{AW}{100}\right) - \dfrac{\text{wt}}{8.33 \times SGa} - \text{wt} + \dfrac{AW}{100}}$$

Where x = additive required, pounds per sack of cement

Wt = required slurry density, lb/gal

SGc = specific gravity of cement

CW = water requirement of cement

AW = water requirement of additive

SGa = specific gravity of additive

$$X = \frac{\left\{\dfrac{17.5 \times 11.207}{3.14}\right\} + (17.5 \times 4.3) - 94 - (8.33 \times 4.3)}{\left(1 + \dfrac{0.34}{100}\right) - \dfrac{17.5}{8.33 \times 5.02} - 17.5 + \dfrac{0.34}{100}}$$

= 15.1 lb of hematite per sk of cement

Question. 209

Number of feet of water in annulus:

Feet = (water added, bbl)/ (Annular capacity, bbl/ft) = 150/ 0.1279 = 1173 ft

Bottomhole pressure decrease

BHP decrease, psi = (12.5 – 8.33) × 0.052 × 1173 = 254 psi

Equivalent mud weight at TD

EMW, ppg = 12.5 – 254/ (0.052 × 10000 ft) = 12.0 ppg

Question. 210

FPC = As x 2500

Where As = pipe wall cross sectional area, sq in

FPC = ($2.875^2 – 2.441^2$ x 0.7854) x 2500 = 4530

Feet of free pipe = (stretch, inch × FPC)/ pull force in thousands of pounds

= 25 × 4530 / 20

= 5663 ft

Question. 211

Difference in pressure gradient, psi/ft:

psi/ft = (11.2ppg - 7.0ppg) x 0.052 = 0.2184

Height, ft = 225psi/0.2184psi/ft = 1030ft

Hence, Less than 1030 ft of spotting fluid should be used to maintain a safety factor that will prevent a kick or blowout.

Question. 212

Pgs = (y / 300 / d) L

Where Pgs = pressure required to break gel strength, psi

y = 10min gel strength of drilling fluid, lb/100 sq ft

d = inside diameter of drill pipe, in.

L = length of drill string, ft

Pgs = (10 / 300 / 4.276) 12,000ft = 93.5psi

Therefore, approximately 94 psi would be required to break circulation.

Question. 213

Barite, sk/100bbl = 1470(14.0 – 12.0)/ (35 –14.0)

N = 140 sk/100 bbl

Question. 215

Starting Volume = 100 (35 – 14.0)/ (35 – 12.0) = 91.3 bbl

Question. 216

Water, bbl = 100 (14.0 – 12.0)/ (12.0 – 8.33)= 54.5 bbl

Question. 217

Formula: $(V_1D_1) + (V_2D_2)$ = VfDf

Where V_1 = volume of fluid 1 (bbl, gal, etc.)

D_1 = density of fluid 1 (ppg, lb/ ft3, etc.)

V_2 = volume of fluid 2 (bbl, gal, etc.)

D_2 = density of fluid 2 (ppg, lb/ ft3, etc.)

V_F = volume of final fluid mix

D_F = density of final fluid mix

$V_1 + V_2 = 300$ .... (1)

(11.0 V1) + (14.0 V2) = (11.5) (300) ...... (2)

Multiply Equation (1) by the density of the lowest mud weight (D1 = 11.0 ppg) and subtract the result from Equation (2):

On solving, we get V2 = 50, and V1 = 300 – 50 = 250 bbl of 11.0 ppg mud

Question. 218

Pressure, psi = (14.2 + 1.0) x 0.052 x 4000ft

Fracture pressure = 3162psi

Also, HPgas = 0.12 x 4000 ft = 480psi

MASP = 3162 – 480 = 2682psi

Question. 219

MASICP = PL – [D x (mud wt2 – mud wtl)] 0.052

Where MASICP = maximum allowable shut-in casing (annulus) pressure, psi

PL = leak-off pressure, psi

D = true vertical depth to casing shoe, ft

Mud wt2 = new mud wt, ppg

Mud wtl = original mud wt, ppg

MASICP = 1040psi – [4000 x (12.5 – 10.0) 0.052]
= 1040psi – 520 = 520psi

Question. 220 - 221

n = 3.32 log (Θ600/ Θ300)

K = Θ600/ 1022n

x = (81600 (K)× n0.387) / (Dh – Dp)n ×MW

AVc = (x)1/(2-n)

GPMc = AVc ×(Dh2 – Dp2)/ 24.5

n = dimensionless

K = dimensionless

X = dimensionless

Θ300 = 300 viscometer dial reading

Θ600 = 600 viscometer dial reading

Dh = hole diameter, in.

Dp = pipe or collar OD, inch

MW = mud weight, ppg

AVc = critical annular velocity, ft/min

GPMc = critical flow rate, gpm

Using above formulae,

n = 0.79, K = 0.2684, x = 1035 t/min, AVc = 310, GPMc = 294 gpm

Question. 222

P = w x cos I

Where P = partial weight available for bit

cos = cosine

I = degrees inclination (angle)

W = total weight of collars

P = 45,000 x cos25= 45,000 x 0.9063 = 40,7841b

Thus, the available weight on bit is 40,7841b.

Question. 223

Hydrostatic Pressure, psi = 0.052 x Mud Weight (ppg) x Depth
= 0.052 x 11x 2500 = 1430 psi

Question. 224

Kill Mud Weight =

$$\frac{10 \times 40}{3000} + 1.2 = 1.3$$

Question. 225

$$\text{Cost / M} = \frac{B + C \times (T + t)}{F}$$

$$\frac{50000 + 10000 \times (50 + 6)}{100} = 61000$$

Question. 226

Minimum flowrate, gpm
= 12.72 x (bit diameter, in.)1.47
= 12.72 x 12.251.47 = 505.87gpm

Hints and Solutions of Important Questions

Chapter 3: Reservoir Engineering

Question. 10

Bulk density of core sample, ρc = 1.9 gm/cc

Density of brine solution, ρb = 1.04 gm/cc

Density of core matrix, ρm = 2.55 gm/cc

Assuming porosity of formation is unit and porosity of formation φ,

So, $\rho c = \varphi x \rho b + (1-\varphi)\rho m$

$\varphi = 0.43 = 43\%$

Question. 11

Pore volume = 200 cu.ft3, Bulk volume = 20 cu.ft3

Porosity of the rock = Pore volume / Bulk volume = 10%

Question. 18

Absolute permeability of rock = 100 mD, relative permeability of to water = 0.40

So, relative permeability of oil = 1 – 0.40 = 0.6

Effective permeability to oil = 0.4 x 100 = 40 mD

Question. 82

Area A = 80 acre, Net pay thickness h = 50 ft; Average porosity φ = 20%

Initial water saturation Swi = 10%, Initial gas saturation Sgi = 10%

Boi = 1.2 rb/stb, Bgi = 0.8rb/mscf

Initial oil saturation Soi = (1-Swi-Sgi) = 0.8

Reservoir bulk volume Vb = 7758 x A(acre) xh(ft) = 31,032,000 bbl

Original oil in place = $\frac{vb\ \phi\ Soi}{Boi}$ 3,546,514 stb

Question. 85-86

Pi = 6000 psi, Pwf = 5500 psi, μ = 0.25 cp, k = 20 md, h = 30 ft

Re = 1000 ft, Rw = 6 in. = 0.5 ft, taking Bo = 1 rb/stb

Oil production rate, q

$$q = \frac{kh\,(Pi - pwf)}{141.2u\ Boi\left[\ln\left(\frac{Re}{Rw}\right) - \frac{3}{4}\right]}$$

1017.8bbl/day

Now Pwf = 3000 psi

q = 6106bbl/day

Question. 89

A =160 acre, h = 20 ft, φ = 0.1, Sw = 0.25, FVF = Bo = 1.0

Recovery factor RF = 0.35

Reservoir pore volume Vb= 7758Ah = 24825600 bbls

Original oil in place (N)

$= \frac{vb\ \phi\ (Swi)}{Boi}$ 1861920 stb

Oil reserve = 0.35N = 651672 bbls

101. Vb= 9418 acre-ft

= 4102489080 cu.ft3

$\varphi = 0.2$, Swi = 0.35, FVF = Bg = 0.004503 ft3/scf

Gas initially on place

$= \frac{vb\ \phi\ (Sgi)}{Bgi}$ 11844 MMscf

Question. 116

$V^B = \frac{\pi\ (2.50)^2}{4}\ 4.00 = 19.6$ cc

Vs = 25 + 50 – (100/50)×25) = 25.0 cc

Ø = 25.0 – 19.6/ 25.0 = 0.216 or 21.6 %

Question. 117

The bulk volume of a core sample was measured by mercury displacement = 25.0 cc.

W_d = 50.25 gm

W_s = 54.50 gm

ρ_l = 0.701 gm/cc

Sample's porosity

V_p = (54.50 – 50.25)/ 0.701 = 6.06 cc

Ø = 6.06/25.0 = 0.242 or 24.2% Grain density

$\rho_s = W_d / V_g = W_d / V_b\ (1 - Ø)$ = 50.25/ 25.0 (1-0.242) = 2.65 gm/cc

Question. 118-121

$Ø = V_b - V_s / V_b$

$V_s = W_s/\rho_s$ = 51.05/2.63 = 19.4 cc

Ø = 23.6 – 19.4/ 23.6 = 0.178 or 17.8%

$S_w = \frac{V_w}{V_p}$ = 1.50/4.2 = 0.357 or 36%

Vo = Wo/ρo =

$\frac{53.50 - (51.05 + 1.50)}{0.850} = 1.12$ cc

So = 1.12 / 4.2 = 0.267 or 27%

Sg = 1 – (So + Sw) = 1 – (0.27 + 0.36)

= 0.37 or 37%

Question. 122

q = 7.07 hk(pe – pw)/ μln(re/ rw)

Where q = bbl/day, h = ft, k = darcys, pe, pw = psi, μ = cp

From this equation; q_o = 350 bbl/day of reservoir oil

$q_{ot} = q_o/B_o$ = 350/1.30 = 2.70 bbl/ day of tank oil

Question. 123

Given: K_{ro} = 0.45 at S_o = 50%

$K_o = (K_{ro})\ (K_o)$ = (0.45) (50) = 22.5 md

Combining B_o in the flow equation,

qot = (7.07)(0.0225)(25)(1000)/ (7.0)(7.64)(1.20) = 62 bbl/day

Question. 124

Given : k_{rw} = 0.23, k_{ro}= 0.23 at S_o = 0.40.

$q_{ot} = q_o/B_o = 7.07\ hk_o(p_e - p_w)/ Bo\mu ln(r_e/r_w)$

Also, $q_w = 7.07\ hk_w(p_e - p_w)/ \mu ln(r_e/r_w)$

Then, water-oil ratio will be

$$q_w/q_{ot} = \frac{k_w\mu_o}{k_o\mu_w} \times B_0 = \frac{k_{rw}\mu_o}{k_{r_o}\mu_w} \times B_0$$

$$\frac{k_w}{k_r} \times B_r \frac{k_{rw} i_r}{n} \times B_r = \frac{(0.23)(7.5)}{(0.23)(0.6)} \times 1.20 = 15$$

Note: $\frac{k_w\mu_o}{k_o\mu_w}$ is called as mobility ratio of water to oil, λw/λo

Question. 125

$q_i = q_i e^{-Dt}$

where, q_i = initial production rate vol/unit time

D = nominal exponential decline rate, 1/time

t = time

q_i = 1130

q = 1050

t = 3 month

$$\text{in}\left(\frac{q}{qi}\right) = -Dt$$

$$D = \text{In}\left(\frac{qi}{q}\right) \Big/ t$$

$$D = \text{in}\left(\frac{1130}{q1050}\right) /3$$

D = 0.22447582b per month

$q = q_i e^{-Dt} = 1130e^{-t-0.02447582(12)}$

q = 842 BOPD after 1 year

Question. 126

$$q = q_i\ (1 + bD_i t)^{\frac{-1}{b}}$$

$$Di = \left[\left(\frac{q}{q_i}\right)^{-b} - 1\right] / bt$$

$$Di = \left[\left(\frac{q}{q_i}\right)^{-0.5} - 1\right] / 0.5\ (3)$$

$D_i = 0.02493067$

$$q = q_i\ (1 + bD_i t)^{\frac{-1}{b}}$$

$$q = 1130\ (1 + 0.5\ (0.02493067)\ 12)^{\frac{-1}{b}}$$

q = 855.1 BOPD

Question. 127

$$q = {qi}\big/(1 + Di\ t)$$

$$D_i = {(qi / q - 1)}\big/t$$

Di = 0.02539683

q = 1130/(1+0.02539683(12))

q = 866.1BOPD

Question. 205

$$S = \left[\frac{k}{k_{skin}} - 1\right] \text{In}\left(\frac{r_{skin}}{r_w}\right) =$$

$$\left[\frac{60}{20} - 1\right] \text{In}\left(\frac{2}{0.25}\right) = 4.6$$

Question. 206

The declining rateis given by

$$\frac{dp}{dt} = -\frac{0.23396q}{c_t \pi r_e^2 h \text{ø}} = \frac{-0.23396q}{c_t\ Ah\text{ø}}$$

$q = Q_0 B_0 = (1200)\ (1.3) = 1560$ bbl/day

$$\therefore\ -4.655 = \frac{0.23396\ (1560)}{(12 \times 10^{-6})\ (A)\ (25)\ (0.15)}$$

A = 1742400ft²

Hints and Solutions of Important Questions

Chapter 4: Petroleum Production Operation-1

Question. 37

PDET – Pressure required to detonate firing head, psi

PCT – Coiled tubing pressure, psi

PCTHYD – Hydrostatic pressure of coiled tubing column, psi

PCTHYD – Reservoir depth X Specific weight of well bore fluid

PDET = PCT + PCTHYD

= 500 + (8000 X 0.052 X 8.5)

= 4035 psi

Therefore absolute pressure of 4035 psi required for detonation of hydraulic firing head.

Question. 38

Tubing pressure required for shear release of hydraulic disconnect

PCT – Coiled Tubing Pressure, psi

PDISC – Differential pressure to shear the disconnect, psi

PCTHYD – Hydrostatic pressure of coiled tubing column, psi

PTHYD – Hydrostatic pressure of tubing column, psi

PCT = PDISC – PCTHYD – PTHYD

PDISC = 1.2 X PDET

= 1.2 X 4035

= 4840 psi

PCT = 4840 – [(8000 X 0.052 X 8.5)–(2595 X 0.05 X 8.5)]

= 2450 psi

Question. 53

Average fracture width (PKN model) = 0.3 [qi μ (1-υ) χf /G]$^{1/4}$ (πϒ/4)

Where qi = Injection rate

μ = Apparent viscosity

υ = Poisson ratio

χf = Fracture length

G = Elastic shear modulus

ϒ = 0.75

E = Youngs modulus

G = E/[2(1+ υ)]

= (4 X 10^{6})/ [2(1+0.25)]

= 1.6 X 10^{6} psi

ϖ = 0.3 [40 X 100 (1-0.25) X 1000/ (1.6 X 10^{6})]$^{1/4}$ (πX

0.75/ 4)
= 0.35 X 0.59
= 0.21 in

0.35 is maximum fracture width and 0.21 is average fracture width.

Question. 59

Fracture gradient (at current reservoir pressure), gf = α + (1- α) x (reservoir pressure/depth)

To find the value of α we uses the above equation with initial reservoir pressure of 3075 psi and fracture gradient of 0.7 psi/ ft.

0.7 = α + (1– α) X 3075/7500

Therefore α = 0.49

Approximately fracture gradient at current reservoir pressure is

gf = 0.49 + (1 – 0.49) X (2000/7500) psi/ft
= 0.63 psi/ft

Question. 60

The maximum injection rate

= i_{max}

= [4.917 X 10^{-6} $K_{av}h_n$ (gf X depth – reservoir pressure)]/ [μ ln (r_e/r_w)]

Where K_{av} = Average permeability

h_n = Perforation interval

r_e = Drainage radius

g_f = Fracture gradient

r_w = Wellbore radius

μ = Acid viscosity

i_{ma} = (4.917 X 10^{-6} X 200) [(7500 X 0.63 –2000)/ 0.4 ln (660/0.25)]

= 0.85 bbl/min

Maximum rate = 0.9 X 0.85 bbl/ mi

= 0.77 bbl/min

Question. 61

Maximum surface injection pressure, P_{max} = (gf – Acid hydrostatic gradient) X Depth

= (0.63-0.43) X 7500
=1500 psi

Question. 62

Fracture gradient at current reservoir pressure = ?

0.7 = α + (1- α) X 2000/5000 psi/ ft

α = 0.5

Therefore, gf = 0.5 + 0.5 X (1000/5000) psi/ft
= 0.6 psi/ft

Question. 63

The maximum injection rate (in practice)

i_{max} = [4.917 X 10^{-6} $K_{av}h_n$ (gf X depth – reservoir pressure)]/ [μ ln (r_e/r_w)]

= [4.917 X 10^{-6} X 50 X10 (5000 X 0.6 – 1000)]/ [0.78 X ln (660/0.25)]
= 0.80

In practice i_{max} = 0.9 X 0.80 = 0.72 bbl/min

Question. 64

Maximum surface injection pressure (P_{max}) = (0.6-0.47) X 5000

= 650 psi

0.47 psi/ft is hydrostatic gradient of 15%HCl

Question. 175

Cycle frequency for intermittent lift operation, N = 1440/[3 X (D_v/1000)] , D_v is Valve depth

= 1440/[3 X (4500/1000)]

= 106.66 =107 cycles/day

Question. 179

Absolute open flow potential, q_o

= [7.08 x K_o x h (P_R - P_{wf})]/ [$B_o\mu_o$ ln (r_e/r_w)]

= [7.08 x 30 x 40 x (3000-0)]/ [1000 x 1.2 x 0.8 ln (1489.83/1.20)]

= 3644 stb/day

Question. 179 - 181

Drainage radius, re = (A× 43560/π) ft

Wellbore radius, r_w = 0.51 ft

Applying Darcy's law for radial flow,

$$q_0 \frac{7.08 \times 10\text{-}3 \times kh\ (Pr - Pwf)}{\mu_0 B_0 \left[\ln (r_e/r_w) - 0.75\right]}$$

Absolute open-flow potential,

$$q_i = \frac{7.08 \times 10\text{-}3 \times 30 \times 40\ (3000 - 0)}{(0.8 \times 1.2)\left[\ln (1490/0.51) - 0.75\right]}$$

= 3672 stb/day

Productivity Index =

$$\frac{q}{(Pr - Pwf)}$$

$$\frac{7.08 \times 10 - 3 \times kh}{\mu_0 B_0\left[\ln(re/rw) - 0.75\right]} = 1.22$$

(bbl/psi-day)

Question. 182

The productivity index Jn can be obtained by dividing qh by ΔP:

J_h = q/ΔP

$$\frac{0.536kh\ \Delta P}{B\mu}\left[1/\ln (4_{reh}/L)\right]$$

$$J_h = \frac{0.007078 \times 75 \times 30}{1.3 \times 0.62}[1/\ln(4 \times 1053/1000)$$

= 13.3 st.bbl/day-psi

= 333.6 m³/(d.MPs)

Question. 183 - 184

$$\Delta P_{skin} = \frac{141.2 \times q\mu B_0}{kh} \times s_t = 141.2 \times (100/20) \times 10 = 1412 \text{ psi}$$

K = 5md pr = 2500 psig

h = 20ft s = – 5,–1, 0,1,5,10,50

μ_o = 1.1 cp B_o = 1.2 RB/STB

Spacing = 80 acres r_w = 0.365 ft

Drainage radius, re =

$$\sqrt{(80 \times 43560)/\eth}$$

= 1053 ft

AOFP =

$$\frac{7.08 \times 10^{-3} \times kh\ (Pr)}{\mu_0 B_0\ \left[\ln(re/rw) - 0.75 + s\right]}$$

$$\frac{7.08 \times 10^{-3} \times 5 \times 20\ (2500)}{(1.1 \times 1.2)\left[\ln\left(\frac{1053}{0.365}\right) - 0.75 + s\right]}$$

$$\frac{1341}{7.97 - 0.75 + s}$$

Question. 180

PI = hk/$B_o\mu_o$

= (40 X 30)/ (.2 X 0.8 X 1000)

= 1.25 stb/psi – day

Question. 181

Production rte at P_{wf} = 100 psig

qo = [7.08 x K_o x h (P_R – P_{wf})]/ [$B_o\mu_o$ ln (r_e/r_w)]

= [7.08 x 30 x 40 x (3000-1000)]/ [1000 x 1.2 x 0.8 ln (1489.83/1.20)]
= 2429.11 stb/day
= 2440 stb/day

Question. 182

The productivity index, P.I = q_n/ Δp

q_n = [(0.007078 K_n h Δp)/Bμ] [1/ ln(4r_{eh}/ L)]

where q_n – flow rate stbbl/day

Δp – pressure drop from the drainage radius to the wellbore , psi

r_{eh} – Horizontal well drainage radius , ft

L – Horizontal well length, ft

K_h – Horizontal permeability, md

B – Formation volume factor, res.bbl/st.bbl

μ – oil viscosity ,cP

P.I = [(0.007078 x 75 x 30)/ (1.34 x 0.62)] [1/ln 4(1053/1000)]
=13.35 st.bbl/D – psi

Question. 183

qo = 100 BPD, Kh/μ = 20 md-ft/ cP, St =2, B_o =1

ΔPskin = 0.87Sm
= 0.87s x 162.5 q_o μ_o B_o / (K_o h)
= 0.87 x 2 x 162.5 x 100 x1 x 1/20
= 1412 psi

Question. 184

Absolute open flow potential,
qo = [7.08 Kh (P_r – P_{wf})]/[$B_o\mu_o$ ln {(r_e/r_w) – 1}]
= [7.08x5x 20 (2500 – 0)]/ [1000 x 1.2 x 1.1 x ln{(1053.20/0.365)-1}]
= 192 stb/day

Question. 186

Average reservoir pressure, Pr = 3000 psig

Q_o = 500 STB/day

Wellbore pressure = 2500 psig

Flowing bottom hole pressure = 1200 psi

Oil production rate=?

q_o/ $(q_o)_{max}$ = 1– 0.20 (P_{wf}/P_r) – 0.80 (P_{wf}/ P_r)^2

500/$(q_o)_{max}$ = 1– 0.20 (2500/3000) – 0.80 (2500/3000)^2
= 1– 0.1666-0.5555

q_{omax} = 500/.2779=1779.20

Therefore, qo/1799.20 = 1 – 0.20 (1200/3000) – 0.8^0 (1200/3000)^2

q° = 1799.20 (1 – 0.08 – 0.128)
= 1424.96 stb/day

Question. 188

Gilbert choke equation is P_1 – D = A q_L F_{gL}B / d_{64}C

Where, P_1 – Upstream pressure, psia

q_L – Liquid flow rate, stb/day

F_{gL} – gas liquid ratio, scf/stb

d_{64} – choke diameter in 64th of an inch

d_{64} = 32 indicates a choke diameter of 0.5 inch

A,B,C,D – Experimentally determined constants

Therefore, 200 –14.7 = 10.0 x 415 x 600$^{0.546}$ / d_{64}1.89

d_{64} = 32 (1/64th of an inch)

Question. 190

q_o = [7.08 Kh (P_r – P_{wf})]/[$B_o\mu_o$ ln {($r_{e/}r_w$)+s}]

640 = 7.08 x 25 x 120 x (4500 –P_{wf}) / [1.3 x = 600/0.5 = 1200 ft

Question. 209

Depth of second gas lift valve = [(525 –1200 x 0.04)/0.5] +1200
= 2154 ft

Question. 210

Depth of third gas lift valve = [(500 – 2154 x 0.04)/0.5] +1200
= 2154 ft

Question. 211

Depth of forth gas lift valve =[(475 – 2982 x 0.04)/0.5] + 2982
= 3694 ft

Question. 212

Depth of fifth gas lift valve = [(450 – 3694 x 0.04)/0.5] + 3694
= 4298 ft

Question. 213

Depth = (Kick off pressure – well head pressure)/ fluid gradient
= 1000 – 120 / 0.40 = 2200 ft

Question. 214

Valve depth (2nd) = [2200 + 1000 – 120 – (0.14 x 2220)] / 0.40
=3630 ft

Question. 215

Valve depth (3rd) = Valve depth 2nd + surface operating pressure 3rd – well head pressure – [(Design gas gradient x valve depth 2nd)/ Fluid gradient]
= 3630 + [(975 -120- 0.14 x 3630)/ 0.4]
= 4497 ft

Question. 216

Valve depth 4th = 4497 + [(950 – 120 – 0.14x 4497)/0.4]
= 4498 ft

Question. 217

Valve depth 5th = 4498 + [(925 –120 – 0.14 x 4498)/ 0.40]
= 5261 ft

Question. 218

Valve depth 6th = 5261 + [(900 – 120 – 0.14 x 5261)/0.40]
= 5370 ft

Question. 222

$J = \dfrac{q}{p - Pwf}$ = 100/ (2000 – 1500)

J = 0.2 STB/psi-D

Question. 223

$$\frac{141.2 JB\mu \left[\left(\frac{r_e}{r_w}\right) - \frac{3}{4} \right]}{h}$$

$$\frac{(141.2)(0.2)(1.5)(0.5)\left[\text{In}\frac{1000}{0.25} - 0.75\right]}{10}$$

= 16 md

Question. 224

$$S = \left(\frac{k}{kj} - 1\right)\left[\text{In}\,\frac{r_e}{r_w} - \frac{3}{4}\right]$$

$$= \left(\frac{50}{16} - 1\right)\left[\text{In}\,\frac{1000}{0.25} - \frac{3}{4}\right]$$

= 16

Question. 225

W_r = 5900×1.814 = 10703 lb

W_f = 0.433 G (LA_p – 0.294W_r)

= 0.433×1.0×(5900×1.227 – 0.294×10703)

= 1772 lb

α = SN^2/70500 = 64×16.5/70500 = 0.247

PPRL = W_f + W_r (1+0.6α)

= 1772 + 10703 (1+0.6 x 0.247) = 14061 lb

Question. 226

MPRL = W_r (1 – 1.4α – 0.127G)

= 10703 (1 – 1.4 x 0.247 – 0.127 ×1.0)

= 5643 lb

Question. 227

C = (PPRL + MPRL) /2

Question. 228

PT = ½(14061×29- 5643×37)/0.93

= 107,000 in – lb

Question. 229

W_r = 5900×1.814 = 10703 lb

W_f = 0.433 G (LA_p – 0.294W_r)

= 0.433×1.0×(5900×1.227 – 0.294×10703)

= 1772 lb

α = SN^2/70500 = 64×16.5/70500

= 0.247

PPRL = W_f + W_r (1+α)

= 1772 + 10703 (1+0.247)

= 15119 lb

Question. 230

W_r = 5900 × 1.814 = 10703 lb

W_f = 0.433 G (LAp – 0.294W_r)

= 0.433 × 1.0 × (5900 ×1.227 – 0.294 × 10703)

= 1772 lb

α = SN^2/70500 = 64×16.5/70500 = 0.247

MPRL = W_r (1 – α – 0.127G)

= 10703 (1 – 0.247 – 0.127 ×1.0)

= 6700 lb

Question. 231

C = (PPRL + MPRL)/2 = (15119 + 6700)/2

= 10910 lb

Question. 232

PT = (15119 – 10910) × 34/ 0.93

= 153900 in-lb

Question. 234

Pump displacement = q/E_v = 250/0.8

Also, Pump displacement = 0.1484 A_pS_pN

Therefore, 0.1484 $A_pS_pN = q/E_v$

N = 16 spm

Question. 235

$$D_{64} = \left(\frac{Aql\ (GLR)^B}{P_1}\right)^{1/c} = \left(\frac{(10)\ (2000)_1\ (500)^{0.546}}{800 - 14.7}\right)^{1/1.89}$$

= 33 (64th of an inch)

Question. 236

Average reservoir pressure, Pr = 2500 psig

Q_o = 350 STB/day

Flowing bottom hole pressure = 2000 psi

$q_o/(q_o)_{max} = 1 - 0.20\ (P_{wf}/P_r) - 0.80\ (P_{wf}/P_r)^2$

$350/(q_o)_{max} = 1 - 0.20\ (2000/2500)$

$- 0.80\ (2000/2500)^2$

Therefore, $q_o/\ q_o)_{max}$ = 1-0.20 $(1850/2500 - 0.80\ (1850/2500)^2$

q_o = 1799.20 (1-0.08-0.128)

= 1067.1 stb/day

Question. 260

Kill Fluid Density = Reservoir Pressure/ (0.052 x Depth) = 2700/(.052x5000)

Hints and Solutions of Important Questions

Chapter 5: Petroleum Production Operation-II

Question. 55

API= (141.5 / SG) – 131.5

Where, API = Degrees API Gravity and SG = Specific Gravity (at 60oF)

Question. 56

API = (141.5 / SG) - 131.5

Where, API = Degrees API Gravity and SG = Specific Gravity (at 60oF)

Question. 227

$$V_1 A_1 = V_2 A_2$$

Question. 240

$$d^2 = 5040\left[\frac{TZQ_{\grave{e}}}{P}\right]\left[\left(\frac{p_g}{p_l - p_g}\right)\frac{C_D}{d_m}\right]^{1/2}$$

$$d^2 = 5040\left[\frac{(520)(0.84)(10)}{1000}\right]\left[\left(\frac{3.71}{51.5 - 3.71}\right)\frac{0.851}{140}\right]^{1/2}$$

Question. 241

$q = 16Q_o \Delta T[0.5(S.G.)_o + (0.1)]$

$q = (16)(2000)(30)[(0.5)(0.825) + 0.1]$

Question. 244 - 245

Shell Thickness

$$t_s = \frac{pD}{2fj + p}$$

f = allowable stress

J = Joint efficiency

D = Diameter

D = Diameter

Header Shell Thickness

$$t_h = \frac{pD}{4fJ}$$

For Shell Thickness

$$p = 10^3\ \frac{KN}{m^2} = 10^6\ N/m^2$$

f = 1.2×10^5 KN/m² = 1.2×10^8 N/m²

D = 0.6 m

J = 1

$$\therefore ts = \frac{1 \times 10^6 \times 0.8}{2 \times 1.2 \times 10^8 + 1 \times 10^6} =$$

$$2.49 \times 10^{-3}\ m$$

Corrosion Allowance = 3 x 10^{-3}m

$\therefore t_s = 5.49 \times 10^{-3}$m

For Header Thickness

$$p = 10^3\ \frac{KN}{m^2} = 10^6\ N/m^2$$

$$\therefore t_h = \frac{2 \times 10^6}{4 \times 1.2 \times 10^8 \times 1} = 0.017m$$

Corrosion Allowance = 3×10^{-3}m

$\therefore t_h = 0.01948$ m

Question. 246 - 250

$$\text{LMTD} = \frac{(105-40)-(50-25)}{\ln\frac{105-40}{50-25}} = 41.86^{\circ}\text{C}$$

True Temperature Difference

$= F_t \times \text{LMTD} =$

$0.85 \times 41.86 = 35.58^{\circ}$C

Heat transfer area for 1 – 2 exchanger A

Rate of heat transfer = Rate of heat lost organic fluid

$= 10000 \times 2.84 \times (105-50)$

$$\frac{1562000\text{KJ}}{\text{hr}} = 433.0\ \text{KW}$$

$$A = \frac{Q}{U\Delta T_{corr}} =$$

$$\frac{433.9 \times 10^3}{600 \times 35.58} = 20.32\ \text{m}^2$$

LMTD =

$$\frac{(54\ 20)-(54-36)}{\ln\frac{54-20}{54-36}} = 25.16^{\circ}\text{C}$$

Hints and Solutions of Important Questions

Chapter 6: Offshore Drilling and Production Practices

Question. 91

T nominal= (weight of riser in water+ weight of drilling fluid in water) x 1.20.

The weight of 20 inch O.D. riser with 1/2inch wall in water is equal to 146 Ibs. per foot

Drilling fluid weight = 181bs. per gallon; weight in water equals 147 Ibs. per foot

Nominal tension in 800 foot water depth = (800 ft.) (147 Ibs. per ft. + 146 Ibs. per ft.) x (1.20) = 281,280 Ibs.

Question. 104

(Heave (ft)x Pi (k))/(period (sec))=

$$\text{Maximum velocity}\left(\frac{\text{ft}}{\text{sec}}\right)$$

heave = 16 ft.; period = 16 sec

Question. 131 - 134

Weightof 20 inch OD riser x 1/2inch wall on water =-33 lbs./ ft.

18lbs./gal. mud = 264 lbs./ft. between ID of riser and ODof drill pipe

Nominal tension for 800 ft. of water, Tnom = (800) (264-33) (1.20) = 222 kips

Operating Tension =

$$\frac{\text{Maximum Tension}}{\text{Safety Diviser}} = \text{OT}$$

Minimum Tension = Riser system weight plus 5%

Safety Diviser =

$$\frac{\text{Number of tension units less one}}{\text{Number of Tension units}} = \text{S. D.}$$

Therefore

S.D.for a 4 tensioner system =

$$\frac{4-1}{4} = 0.75.$$

S.D.for a 6 tensioner system =

$$\frac{6-1}{6} = 0.83.$$

$$OT_4 = \frac{(800\text{ ft})\left(231\frac{\text{lbs}}{\text{ft}}\right)(1.05)}{0.75} = 259\text{ kips.}$$

$$OT_6 = \frac{(800\text{ ft})\left(231\frac{\text{lbs}}{\text{ft}}\right)(1.05)}{0.83} = 234\text{ kips.}$$

Hints and Solutions of Important Questions

Chapter 7: Petroleum Formation Evaluation

Question. 38

$$\text{ø} = \text{ø}_o \times e^{-(\text{depth}/a)}$$

$$\therefore \text{ø} = 45 \times e^{-(15000/12000)}$$

$$\therefore \text{ø} = 0.129$$

$\therefore$ Porosity = 12.9%

Question. 94

$$\text{ø} = \frac{\rho_{\text{matrix}} - \rho_{\text{bulk}}}{\rho_{\text{matrix}} - \rho_{\text{fluid}}}$$

$$\text{ø} = \frac{\rho_{\text{matrix}} - \rho_{\text{bulk}}}{\rho_{\text{matrix}} - \rho_{\text{fluid}}}$$

$$\text{ø} = \frac{2.68 - 2.20}{2.68 - 1.2}$$

$$= 0.324$$

$\therefore$ Porosity = 32%

Question. 96

$$\text{ø}_s = \left(\frac{\Delta t - \Delta t_{ma}}{\Delta t_p - \Delta t_{ma}}\right) \times \frac{1}{C_p}$$

$$\text{ø}_s = \left(\frac{90 - 51}{189 - 51}\right) \times \frac{1}{1.2}$$

$$\text{ø}_s = 0.246$$

$\therefore$ Porosity = 24.6%

Question. 97

$$S_w = \sqrt[n]{\frac{R_w F}{R_t}}$$

$$S_w = \sqrt[2.3]{\frac{(0.05)\,(26)}{10}}$$

$$S_w = \sqrt[2.3]{\frac{(0.05)\,(26)}{10}}$$

$$S_w = 0.4118$$

$\therefore$ Porosity = 41%

Hints and Solutions of Important Questions

Chapter 8: Oil and Gas Well Testing

Question. 67

$P(r,t)= P_i +$

$$\left[\frac{70.6Q_o \mu_o B_o}{kh}\right] \times Ei \left[\frac{-948\emptyset\mu_o c_t r^2}{kt}\right]$$

$P(r,t) = 4000+$

$$\left[\frac{70.6\,(300)\,(1.5)\,(1.25)}{(60)\,(15)}\right] \times$$

$$Ei\left[\frac{-948\,(1.5)\,(12\times10^{-6})\,r^2}{(60)\,(t)}\right]$$

= 3785 psi

Question. 68

Dimensionless time t_D

$$t_D = \frac{0.000264kt}{\emptyset\mu_o\, c_t r_w^2} = 93,86667$$

$$t_D = \frac{0.000264\,(60)\,(1)}{(0.15)\,(1.5)\,(12\times10^{-6})\,(0.25)^2}$$

$= 93,866.67$

Since tD> 100 calculate the dimensionless pressure drop function

$p_D = 0.5\ [\ln(t_D) + 0.80907]$ $=6.1294$

$p_D = 0.5\ [\ln(93,866.67) + 0.80907]$ $=6.1294$

The bottom hole pressure after 1 hr.

$$p(r,t)=p_i - \frac{Q_o\mu_o B_o}{0.00708kh}(p_D)$$

$p(0.25,1)=4000 -$

$$\left[\frac{(300)\,(1.25)\,(1.5)}{0.00708\,(60)\,(15)}\right]$$

$(6.1294)=3459$ psi

Question. 69

Wellbore storage factor caused by fluid expansion is given by

$C_{FE}=V_{wb}\, C_{wb}$

Where

C_{FE} is the Well bore storage due to expansion of fluid

V_{wb} is volume of fluid in the well bore

C_{wb} is the compressibility of the fluid

$\therefore C_{FE}= (1800)(10\times10^{-6})=0.0018$ bbl/psi

Question. 70

Wellbore storage factor caused by falling fluid level is given by

$$C_{fl}= \frac{144A_b}{5.615_\rho}$$

$$C_{fl}= \frac{144\,(0.2995)}{(5.615)\,(45)} = 0.1707 \text{ bbl/day}$$

Question. 71

Total wellbore storage coefficient

$$C=C_{FE}+C_{fl}$$

$C=0.0018+0.1707=0.1725$ bbl/day

Question. 72

The time required for well bore storage influence to end from:

$$t=\frac{(200000+12000_s)c\mu}{kh}$$

$$=\frac{(200000+0)(0.17250)(2)}{(30)(50)}$$

= 46 hours

Question. 73

The average permeability is given by

$$k=\frac{162.6\,Q_o\,B_o\,\mu_o}{mh}$$

$$\frac{(162.6)(49000)(1.55)(0.22)}{(40)(482)}$$

= 12.8 md

Question. 74

Skin Factor is given by

$$s=1.151\left[\frac{p_{1\,hr}-pwf(\Delta t=0)}{m}-10\,g\left(\frac{k}{\O\mu\;c_t\;r_w^2}\right)+3.23\right]$$

=1.151

$$\left[\frac{3266-2761}{40}-\text{lo g}\left(\frac{12.8}{(0.09)(0.20)(22.6\times10^{-8})(0.354)^2}\right)+3.23\right]$$

=8.6

Question. 75

Effective radius

$$r_s=\sqrt{\frac{40\times43560}{\pi}}=745\text{ ft}$$

Question. 76

$$k_{impr}=\frac{q_o\,\mu_o\,\beta_o\,\text{In}\left(\frac{r_s}{r_w}\right)}{0.00708\,h\,(ps-pw-(\Delta p)_{skin})}$$

$$=\frac{800\times2.0\times1.255\,\text{In}\left(\frac{745}{0.27}\right)}{0.00708\times45\,(1350-800-150)}=$$

124.83 mD

Question. 77

$$q_i=\frac{0.00708kh\left[p_s-p_w-(\Delta p)_{skin}\right]}{\mu_o\,\beta_o\,\text{In}\left(\frac{r_s}{r_w}\right)}$$

$$75=\frac{0.00708\times10\times15\,(2200-950-(\Delta P)_{skin})}{0.95\times1.24\,\text{In}\left(\frac{750}{0.39}\right)}$$

$(\Delta p)_{skin}$=140.32 psi

Question. 78

$$q_i=\frac{0.00708\,kh\left[p_s-p_w\right]}{\mu_o\,\beta_o\,\text{In}\left(\frac{r_s}{r_w}\right)}$$

$$=\frac{0.00708\times10\times15\,(2200-950)}{0.95\times1.24\,\text{In}\left(\frac{750}{0.39}\right)}$$

=149.02 stb/day

Question. 79

$$149.02=\frac{0.00708\times10\times15\,(2200-950-(\Delta p)skin)}{0.95\times1.24\,\text{In}\left(\frac{750}{5.5}\right)}$$

$(\Delta p)_{skin}$=464 psi

Hints and Solutions of Important Questions

Chapter 10: Enhanced Oil Recovery

Question. 79

Volume of N_2 injected = Volume of produced gas × Percentage of N_2 gas in produced gas

$= Vg \times \%N_{2P}$

$= 190 \times 84.2\ \%$

$= 160$ SCF

Volume of oxygen gas injected = Volume of N_2 injected × (% of O_2 in injected air/ % of N2 in injected air)

$= (V_g \times \%\ N_{2P}) \times (\%O_{2I}\ /\ \%\ N_{2I})$

$= 160 \times (21/79)$

$= 42.5$ SCF

Question. 80

Volume of CO_2, CO and unreacted O_2 produced in the laboratory

Volume of CO_2 produced = (Volume of produced gas) × (% of CO_2 in produced gas)

$= 190 \times 11.7\%$

$= 22.23$ SCF

Volume of CO produced = volume of produced gas × % of CO in produced gas

$= 190 \times 3.0\ \%$

$= 5.7$ SCF

Volume of unreacted O_2 = Volume of produced gas × (% of O_2 in produced gas)

$= 190 \times 1.1\%$

$= 2.09$ SCF

Question. 81

Carbon in fuel burned (Wc, lb) = [(Volume of CO_2 produced + Volume of CO produced)] × (12/379)

$= (22.23 + 5.7) \times (12/379)$

$= 0.88433$ lb.

Hydrogen in fuel burned (WH), lb = 2[Volume of oxygen injected – volume of unreacted oxygen produced – volume of carbon dioxide produced – (0.5 × volume of carbon dioxide produced)] × (2/379)

$= 2\ [42.5 - 2.09 - 22.23 - 0.5 \times 5.7] \times (2/379)$

$= 0.1618$ lb

Weight of water formed in pounds (W_w) = 2 [Volume of oxygen injected – volume of unreacted oxygen produced – volume of carbon dioxide produced – 0.5 × volume of carbon dioxide produced] × (18/379)

$= 2\ [42.5 - 2.09 - 22.23 - 0.5 \times 5.7] \times (18/379)$

$= 1.46$ lbs of water

Question. 82

Weight of total fuel consumed in the laboratory run (W_F) = Weight of carbon in fuel burned + weight of hydrogen in fuel burned

= 0.88433 + 0.1618 = 1.04613 lb

Question. 83

D – Combustion tube internal diameter

L – Length of burned pack

Fuel consumed / ft_3 of sand pack burned (W_P) = (4 × wt. of total fuel consumed) × (π × D_2 × L)

= (4 × 1.04613) / (3.13 × 0.333^2 × 6.0)

= 2.002 lb / ft^2

Fuel consumed / acre-ft of reservoir, W_R = 43560 × Fuel used/ ft_3 × [(1 – porosity of reservoir) / (1– porosity of pack)

= 43560 × 2.002 ×[(1– 0.18) / (1 – 0.35)]

= 110015.14 lbs/ acre-ft

Question. 84

Air injected / acre–ft of reservoir burned = 43560 × Air scf/ft^3

= [4 × volume of air × (1– porosity of reservoir)]/ [(π × D^2 × L) (1– porosity of pack)]

= 4 × (Volume of N_2 + Volume of O_2) × [(1– porosity of reservoir)/(1– porosity of pack)]

= [{4 × (160 + 42.5) × (1-0.18)} / {3.14 × 0.333^2 × 6.0 × (1 – 0.35)}]

= 489.1059 scf of air/ ft^3 of reservoir

Therefore, Air injected/acre – ft = 489.1059 × 43560

= 21305453.33 scf

= 21.30 MMSCF

Question. 85

Considering 62.6% horizontal sweep efficiency at infinite mobility ratio

Air injected/ acre-ft of pattern

= 0.626 × 21.30

= 13.33 MMSCF/acre – ft

Total air required for 5 – spot pattern = 0.626 × Air injected/ acre-ft × Fuel consumed/ft^3 × Formation thickness × 3302

= 0.6262 × 489.0 × 2.00 × 80 × 10^{-6} × 330^2

= 5300 MMSCF

Question. 86

Air flux for a minimum burning front advance of 0.125 ft/day = Air injected/ acre – ft × 0.125

= 489.0 × 0.125

= 61.6 scf/ft^2– day